MATHLETICS

MATHLETICS

How Gamblers, Managers, and Sports Enthusiasts Use
Mathematics in Baseball, Basketball, and Football

WAYNE WINSTON

PRINCETON UNIVERSITY PRESS   PRINCETON AND OXFORD

Copyright © 2009 by Princeton University Press
Published by Princeton University Press, 41 William Street,
Princeton, New Jersey 08540

In the United Kingdom: Princeton University Press, 6 Oxford Street,
Woodstock, Oxfordshire OX20 1TW

Library of Congress Cataloging-in-Publication Data
Winston, Wayne L.
Mathletics : how gamblers, managers, and sports enthusiasts use
mathematics in baseball, basketball, and football / Wayne Winston.
        p.   cm.
Includes bibliographical references and index.
ISBN 978-0-691-13913-5 (hardcover : alk. paper)
    1. Sports—Mathematics.   I. Title.
        GV706.8.W56 2009
    796.0151—dc22   2008051678

British Library Cataloging-in-Publication Data is available

This book has been composed in ITC Galliard

Printed on acid-free paper. ∞

press.princeton.edu

Printed in the United States of America

1 3 5 7 9 10 8 6 4 2

To Gregory, Jennifer, and Vivian

CONTENTS

PREFACE

If you have picked up this book you surely love sports and you probably like math. You may have read Michael Lewis's great book *Moneyball*, which describes how the Oakland A's used mathematical analysis to help them compete successfully with the New York Yankees even though the average annual payroll for the A's is less than 40 percent of that of the Yankees. After reading *Moneyball*, you might have been curious about how the math models described in the book actually work. You may have heard how a former night watchman, Bill James, revolutionized the way baseball professionals evaluate players. You probably want to know exactly how James and other "sabermetricians" used mathematics to change the way hitters, pitchers, and fielders are evaluated. You might have heard about the analysis of Berkeley economic professor David Romer that showed that NFL teams should rarely punt on fourth down. How did Romer use mathematics to come up with his controversial conclusion? You might have heard how Mark Cuban used math models (and his incredible business savvy) to revitalize the moribund Dallas Mavericks franchise. What mathematical models does Cuban use to evaluate NBA players and lineups? Maybe you bet once in a while on NFL games and wonder whether math can help you do better financially. How can math determine the true probability of a team winning a game, winning the NCAA tournament, or just covering the point spread? Maybe you think the NBA could have used math to spot Tim Donaghy's game fixing before being informed about it by the FBI. This book will show you how a statistical analysis would have "red flagged" Donaghy as a potential fixer.

If *Moneyball* or day-to-day sports viewing has piqued your interest in how mathematics is used (or can be used) to make decisions in sports and sports gambling, this book is for you. I hope when you finish reading the book you will love math almost as much as you love sports.

To date there has been no book that explains how the people running Major League Baseball, basketball, and football teams and Las Vegas sports bookies use math. The goal of *Mathletics* is to demonstrate how simple arithmetic, probability theory, and statistics can be combined with a large dose of common sense to better evaluate players and game strategy in America's major sports. I will also show how math can be used to rank sports teams and evaluate sports bets.

Throughout the book you will see references to Excel files (e.g., Standings.xls). These files may be downloaded from the book's Web site, http://www.waynewinston.edu).

ACKNOWLEDGMENTS

I would like to acknowledge George Nemhauser of Georgia Tech, Michael Magazine of the University of Cincinnati, and an anonymous reviewer for their extremely helpful suggestions. Most of all, I would like to recognize my best friend and sports handicapper, Jeff Sagarin. My discussions with Jeff about sports and mathematics have always been stimulating, and this book would not be one-tenth as good if I did not know Jeff.

Thanks to my editor, Vickie Kearn, for her unwavering support throughout the project. Also thanks to my outstanding production editor, Debbie Tegarden. Thanks to Jenn Backer for her great copyediting of the manuscript. Finally, a special thanks to Teresa Reimers of Microsoft Finance for coming up with the title of the book.

All the math you need to know will be developed as you proceed through the book. When you have completed the book, you should be capable of doing your own mathletics research using the vast amount of data readily available on the Internet. Even if your career does not involve sports, I hope working through the logical analyses described in this book will help you think more logically and analytically about the decisions you make in your own career. I also hope you will watch sporting events with a more analytical perspective. If you enjoy reading this book as much as I enjoyed writing it, you will have a great time. My contact information is given below. I look forward to hearing from you.

Wayne Winston
Kelley School of Business
Bloomington, Indiana

ABBREVIATIONS

2B	Double
3B	Triple
AB	At Bats
BA	Batting Average
BABIP	Batting Average on Balls in Play
BB	Bases on Balls (Walks)
BCS	Bowl Championship Series
BFP	Batters Faced by Pitchers
CS	Caught Stealing
D	Down
DICE	Defense-Independent Component ERA
DIPS	Defense-Independent Pitching Statistics
DPAR	Defense Adjusted Points above Replacement
DPY/A	Defense-Passing Yards Per Attempt
DRP	Defensive Rebounding Percentage
DRY/A	Defense Rushing Yards Per Attempt
DTO	Defensive Turnover
DTPP	Defensive Turnovers Caused Per Possession
DVOA	Defense Adjusted Value over Average
EFG	Effective Field Goal Percentage
ERA	Earned Run Average
EXTRAFG	Extra Field Goal
FP	Fielding Percentage
FG	Field Goal
FT	Free Throw
FTR	Free Throw Rate
GIDP	Ground into Double Play
GO	Ground Out
HBP	Hit by Pitch
HR	Home Run

IP	Innings Pitched
K	Strikeout
MAD	Mean Absolute Deviation
MLB	Major League Baseball
OBP	On-Base Percentage
OEFG	Opponent's Effective Field Goal Percentage
OFTR	Opponent's Free Throw Rate
ORP	Offensive Rebounding Percentage
OPS	On-Base Plus Slugging
PAP	Pitcher Abuse Points
PENDIF	Penalty Differential
PER	Player Efficiency Rating
PO	Put Out
PORP	Points over Replacement Player
PRESSURE TD	Pressure Touchdown
PY/A	Passing Yards Per Attempt
QB	Quarterback
RET TD	Return Touchdown
RF	Range Factor
RPI	Ratings Percentage Index
RSQ	R-Squared Value
RY/A	Rushing Yards Per Attempt
SAC	Sacrifice Bunt
SAFE	Spatial Aggregate Fielding Evaluation
SAGWINPOINTS	Number of total points earned by player during a season based on how his game events change his team's probability of winning a game (events that generate a single win will add to a net of +2000 points)
SAGWINDIFF	Sagarin Winning Probability Difference
SB	Stolen Base
SF	Sacrifice Fly
SLG	Slugging Percentage
SS	Shortstop
TB	Total Bases
TD	Touchdown
TO	Turnover
TPP	Turnovers Committed Per Possession
TPZSG	Two-Person Zero Sum Game

VORPP	Value of a Replacement Player Points
WINDIFF	Winning Probability Difference
WINVAL	Winning Value
WOBA	Weighted On-Base Average
WWRT	Wald-Wolfowitz Runs Test
YL	Yard Line (where the ball is spotted at start of a play)
YTG	Yards to Go (for a first down)

PART I

BASEBALL

BASEBALL'S PYTHAGOREAN THEOREM

The more runs a baseball team scores, the more games the team should win. Conversely, the fewer runs a team gives up, the more games the team should win. Bill James, probably the most celebrated advocate of applying mathematics to analysis of Major League Baseball (often called sabermetrics), studied many years of Major League Baseball (MLB) standings and found that the percentage of games won by a baseball team can be well approximated by the formula

$$\frac{\text{runs scored}^2}{\text{runs scored}^2 + \text{runs allowed}^2} = \begin{array}{l}\text{estimate of percentage}\\ \text{of games won.}\end{array} \quad (1)$$

This formula has several desirable properties.

- The predicted win percentage is always between 0 and 1.
- An increase in runs scored increases predicted win percentage.
- A decrease in runs allowed increases predicted win percentage.

Consider a right triangle with a hypotenuse (the longest side) of length c and two other sides of lengths a and b. Recall from high school geometry that the Pythagorean Theorem states that a triangle is a right triangle if and only if $a^2 + b^2 = c^2$. For example, a triangle with sides of lengths 3, 4, and 5 is a right triangle because $3^2 + 4^2 = 5^2$. The fact that equation (1) adds up the squares of two numbers led Bill James to call the relationship described in (1) Baseball's Pythagorean Theorem.

Let's define $R = \dfrac{\text{runs scored}}{\text{runs allowed}}$ as a team's scoring ratio. If we divide the numerator and denominator of (1) by (runs allowed)^2, then the value of the fraction remains unchanged and we may rewrite (1) as equation (1)′.

$$\frac{R^2}{R^2+1} = \text{estimate of percentage of games won.} \qquad (1)'$$

Figure 1.1 shows how well $(1)'$ predicts MLB teams' winning percentages for the 1980–2006 seasons.

For example, the 2006 Detroit Tigers (DET) scored 822 runs and gave up 675 runs. Their scoring ratio was $R = \dfrac{822}{675} = 1.218$. Their predicted win percentage from Baseball's Pythagorean Theorem was $\dfrac{1.218^2}{(1.218)^2 + 1} = .597$.

The 2006 Tigers actually won a fraction of their games, or $\dfrac{95}{162} = .586$.

Thus $(1)'$ was off by 1.1% in predicting the percentage of games won by the Tigers in 2006.

For each team define error in winning percentage prediction as actual winning percentage minus predicted winning percentage. For example, for the 2006 Arizona Diamondbacks (ARI), error $= .469 - .490 = -.021$ and for the 2006 Boston Red Sox (BOS), error $= .531 - .497 = .034$. A positive

	A	B	C	D	E	F	G	H	I	J
1										MAD = 0.020
2										
3	Year	Team	Wins	Losses	Runs scored	Runs allowed	Scoring ratio	Predicted winning %	Actual Winning %	Absolute Error
4	2006	Diamondbacks	76	86	773	788	0.981	0.490	0.469	0.021
5	2006	Braves	79	83	849	805	1.055	0.527	0.488	0.039
6	2006	Orioles	70	92	768	899	0.854	0.422	0.432	0.010
7	2006	Red Sox	86	76	820	825	0.994	0.497	0.531	0.034
8	2006	White Sox	90	72	868	794	1.093	0.544	0.556	0.011
9	2006	Cubs	66	96	716	834	0.859	0.424	0.407	0.017
10	2006	Reds	80	82	749	801	0.935	0.466	0.494	0.027
11	2006	Indians	78	84	870	782	1.113	0.553	0.481	0.072
12	2006	Rockies	76	86	813	812	1.001	0.501	0.469	0.031
13	2006	Tigers	95	67	822	675	1.218	0.597	0.586	0.011
14	2006	Marlins	78	84	758	772	0.982	0.491	0.481	0.009
15	2006	Astros	82	80	735	719	1.022	0.511	0.506	0.005
16	2006	Royals	62	100	757	971	0.780	0.378	0.383	0.005
17	2006	Angels	89	73	766	732	1.046	0.523	0.549	0.027
18	2006	Dodgers	88	74	820	751	1.092	0.544	0.543	0.001
19	2006	Brewers	75	87	730	833	0.876	0.434	0.463	0.029
20	2006	Twins	96	66	801	683	1.173	0.579	0.593	0.014
21	2006	Yankees	97	65	930	767	1.213	0.595	0.599	0.004

Figure 1.1. Baseball's Pythagorean Theorem, 1980–2006. See file Standings.xls.

error means that the team won more games than predicted while a negative error means the team won fewer games than predicted. Column J in figure 1.1 computes the absolute value of the prediction error for each team. Recall that the absolute value of a number is simply the distance of the number from 0. That is, $|5| = |-5| = 5$. The absolute prediction errors for each team were averaged to obtain a measure of how well the predicted win percentages fit the actual team winning percentages. The average of absolute forecasting errors is called the MAD (Mean Absolute Deviation).[1] For this data set, the predicted winning percentages of the Pythagorean Theorem were off by an average of 2% per team (cell J1).

Instead of blindly assuming winning percentage can be approximated by using the square of the scoring ratio, perhaps we should try a formula to predict winning percentage, such as

$$\frac{R^{exp}}{R^{exp} + 1}. \tag{2}$$

If we vary exp (exponent) in (2) we can make (2) better fit the actual dependence of winning percentage on scoring ratio for different sports. For baseball, we will allow exp in (2) to vary between 1 and 3. Of course, exp = 2 reduces to the Pythagorean Theorem.

Figure 1.2 shows how MAD changes as we vary exp between 1 and 3.[2] We see that indeed exp = 1.9 yields the smallest MAD (1.96%). An exp value of 2 is almost as good (MAD of 1.97%), so for simplicity we will stick with Bill James's view that exp = 2. Therefore, exp = 2 (or 1.9) yields the best forecasts if we use an equation of form (2). Of course, there might be another equation that predicts winning percentage better than the Pythagorean Theorem from runs scored and allowed. The Pythagorean Theorem is simple and intuitive, however, and works very well. After all, we are off in predicting team wins by an average of $162 \times .02$, which is approximately three wins per team. Therefore, I see no reason to look for a more complicated (albeit slightly more accurate) model.

---

[1] The actual errors were not simply averaged because averaging positive and negative errors would result in positive and negative errors canceling out. For example, if one team wins 5% more games than (1)′ predicts and another team wins 5% fewer games than (1)′ predicts, the average of the errors is 0 but the average of the absolute errors is 5%. Of course, in this simple situation estimating the average error as 5% is correct while estimating the average error as 0% is nonsensical.

[2] See the chapter appendix for an explanation of how Excel's great Data Table feature was used to determine how MAD changes as exp varied between 1 and 3.

	N	O
2		**EXP**
3		2
4	**Variation of MAD as Exp changes**	
5		**MAD**
6	**Exp**	0.0197
7	1.0	0.0318
8	1.1	0.0297
9	1.2	0.0277
10	1.3	0.0259
11	1.4	0.0243
12	1.5	0.0228
13	1.6	0.0216
14	1.7	0.0206
15	1.8	0.0200
16	1.9	0.0196
17	2.0	0.0197
18	2.1	0.0200
19	2.2	0.0207
20	2.3	0.0216
21	2.4	0.0228
22	2.5	0.0243
23	2.6	0.0260
24	2.7	0.0278
25	2.8	0.0298
26	2.9	0.0318
27	3.0	0.0339

Figure 1.2. Dependence of Pythagorean Theorem accuracy on exponent. See file Standings.xls.

How Well Does the Pythagorean Theorem Forecast?

To test the utility of the Pythagorean Theorem (or any prediction model), we should check how well it forecasts the future. I compared the Pythagorean Theorem's forecast for each MLB playoff series (1980–2007) against a prediction based just on games won. For each playoff series the Pythagorean method would predict the winner to be the team with the higher scoring ratio, while the "games won" approach simply predicts the winner of a playoff series to be the team that won more games. We found that the Pythagorean approach correctly predicted 57 of 106 playoff series (53.8%) while the "games won" approach correctly predicted the winner of only 50% (50 out of 100) of playoff series.[3] The reader is prob-

---

[3] In six playoff series the opposing teams had identical win-loss records so the "Games Won" approach could not make a prediction.

ably disappointed that even the Pythagorean method only correctly forecasts the outcome of less than 54% of baseball playoff series. I believe that the regular season is a relatively poor predictor of the playoffs in baseball because a team's regular season record depends greatly on the performance of five starting pitchers. During the playoffs teams only use three or four starting pitchers, so much of the regular season data (games involving the fourth and fifth starting pitchers) are not relevant for predicting the outcome of the playoffs.

For anecdotal evidence of how the Pythagorean Theorem forecasts the future performance of a team better than a team's win-loss record, consider the case of the 2005 Washington Nationals. On July 4, 2005, the Nationals were in first place with a record of 50–32. If we extrapolate this winning percentage we would have predicted a final record of 99–63. On July 4, 2005, the Nationals scoring ratio was .991. On July 4, 2005, (1)' would have predicted a final record of 80–82. Sure enough, the poor Nationals finished 81–81.

The Importance of the Pythagorean Theorem

Baseball's Pythagorean Theorem is also important because it allows us to determine how many extra wins (or losses) will result from a trade. Suppose a team has scored 850 runs during a season and has given up 800 runs. Suppose we trade a shortstop (Joe) who "created"[4] 150 runs for a shortstop (Greg) who created 170 runs in the same number of plate appearances. This trade will cause the team (all other things being equal) to score 20 more runs $(170 - 150 = 20)$. Before the trade, $R = \dfrac{850}{800} = 1.0625$, and we would predict the team to have won $\dfrac{162(1.0625)^2}{1 + (1.0625)^2} = 85.9$ games. After the trade, $R = \dfrac{870}{800} = 1.0875$, and we would predict the team to win $\dfrac{162(1.0875)^2}{1 + (1.0875)^2} = 87.8$ games. Therefore, we estimate the trade makes our team 1.9 games better $(87.8 - 85.9 = 1.9)$. In chapter 9, we will see how the Pythagorean Theorem can be used to help determine fair salaries for MLB players.

[4] In chapters 2–4 we will explain in detail how to determine how many runs a hitter creates.

Football and Basketball "Pythagorean Theorems"

Does the Pythagorean Theorem hold for football and basketball? Daryl Morey, the general manager for the Houston Rockets, has shown that for the NFL, equation (2) with exp = 2.37 gives the most accurate predictions for winning percentage while for the NBA, equation (2) with exp = 13.91 gives the most accurate predictions for winning percentage. Figure 1.3 gives the predicted and actual winning percentages for the NFL for the 2006 season, while figure 1.4 gives the predicted and actual winning percentages for the NBA for the 2006–7 season.

For the 2005–7 NFL seasons, MAD was minimized by exp = 2.7. Exp = 2.7 yielded a MAD of 5.9%, while Morey's exp = 2.37 yielded a MAD of 6.1%. For the 2004–7 NBA seasons, exp = 15.4 best fit actual winning percentages. MAD for these seasons was 3.36% for exp = 15.4 and 3.40% for exp = 13.91. Since Morey's values of exp are very close in accuracy to the values we found from recent seasons we will stick with Morey's values of exp.

These predicted winning percentages are based on regular season data. Therefore, we could look at teams that performed much better than expected during the regular season and predict that "luck would catch up

	B	C	D	E	F	G	H	I	J	K	L	M	N
3			exp = 2.4							MAD = 0.061497			
4													
5	Year	Team	Wins	Losses	Points for	Points against	Ratio	Predicted winning %	Annual winning %	abserr		exp	MAD
6	2007	N.E. Patriots	16	0	589	274	2.149635	0.859815262	1	0.140185			0.061497
7	2007	B. Bills	7	9	252	354	0.711864	0.308853076	0.4375	0.128647		1.5	0.08419
8	2007	N.Y. Jets	4	12	268	355	0.75493	0.339330307	0.25	0.08933		1.6	0.080449
9	2007	M.Dolphins	1	15	267	437	0.610984	0.237277785	0.625	0.174778		1.7	0.077006
10	2007	C. Browns	10	6	402	382	1.052356	0.530199349	0.625	0.094801		1.8	0.073795
11	2007	P. Steelers	10	6	393	269	1.460967	0.710633507	0.625	0.085634		1.9	0.070675
12	2007	C. Bengals	7	9	380	385	0.987013	0.492255411	0.4375	0.054755		2	0.068155
13	2007	B. Ravens	5	11	275	384	0.716146	0.311894893	0.3125	0.000605		2.1	0.06588
14	2007	I. Colts	13	3	450	262	1.717557	0.782779877	0.8125	0.02972		2.2	0.064002
15	2007	J. Jaguars	11	5	411	304	1.351974	0.67144112	0.6875	0.016059		2.3	0.062394
16	2007	T. Titans	10	6	301	297	1.013468	0.507925876	0.625	0.117074		2.4	0.061216
17	2007	H. Texans	8	8	379	384	0.986979	0.492235113	0.5	0.007765		2.5	0.060312
18	2007	S.D. Chargers	11	5	412	284	1.450704	0.707186057	0.6875	0.019686		2.6	0.059554
19	2007	D. Broncos	7	9	320	409	0.782396	0.35856816	0.4375	0.078932	best!	2.7	0.059456
20	2007	O. Raiders	4	12	283	398	0.711055	0.308278013	0.25	0.058278		2.8	0.059828
21	2007	K.C. Chiefs	4	12	226	335	0.674627	0.282352662	0.25	0.032353		2.9	0.060934
22	2007	D. Cowboys	13	3	455	325	1.4	0.689426435	0.8125	0.123074		3	0.062411
23	2007	N.Y. Giants	10	6	373	351	1.062678	0.535957197	0.625	0.089043		3.4	0.063891

Figure 1.3. Predicted NFL winning percentages. Exp = 2.4. See file Sportshw1.xls.

	E	F	G	H	I	J	K
37	2006–2007 NBA						MAD = 0.05
38							
39	**Team**	**PF**	**PA**	**Ratio**	**Predicted Win %**	**Actual Win %**	**Abs. Error**
40	Phoenix Suns	110.2	102.9	1.07	0.722	0.744	0.022
41	Golden State Warriors	106.5	106.9	1.00	0.487	0.512	0.025
42	Denver Nuggets	105.4	103.7	1.02	0.556	0.549	0.008
43	Washington Wizards	104.3	104.9	0.99	0.480	0.500	0.020
44	L.A. Lakers	103.3	103.4	1.00	0.497	0.512	0.016
45	Memphis Grizzlies	101.6	106.7	0.95	0.336	0.268	0.068
46	Utah Jazz	101.5	98.6	1.03	0.599	0.622	0.022
47	Sacramento Kings	101.3	103.1	0.98	0.439	0.395	0.044
48	Dallas Mavericks	100	92.8	1.08	0.739	0.817	0.078
49	Milwaukee Bucks	99.7	104	0.96	0.357	0.341	0.016
50	Toronto Raptors	99.5	98.5	1.01	0.535	0.573	0.038
51	Seattle Supersonics	99.1	102	0.97	0.401	0.378	0.023
52	Chicago Bulls	98.8	93.8	1.05	0.673	0.598	0.076
53	San Antonio Spurs	98.5	90.1	1.09	0.776	0.707	0.068
54	New Jersey Nets	97.6	98.3	0.99	0.475	0.500	0.025
55	New York Knicks	97.5	100.3	0.97	0.403	0.402	0.000
56	Houston Rockets	97	92.1	1.05	0.673	0.634	0.039
57	Charlotte Bobcats	96.9	100.6	0.96	0.373	0.402	0.030
58	Cleveland Cavaliers	96.8	92.9	1.04	0.639	0.610	0.029
59	Minnesota Timberwolves	96.1	99.7	0.96	0.375	0.395	0.020
60	Detroit Pistons	96	91.8	1.05	0.651	0.646	0.004
61	Boston Celtics	95.8	99.2	0.97	0.381	0.293	0.088
62	Indiana Pacers	95.6	98	0.98	0.415	0.427	0.012
63	L.A. Clippers	95.6	96.1	0.99	0.482	0.952	0.471
64	New Orleans Hornets	95.5	97.1	0.98	0.442	0.476	0.033
65	Philadelphia 76ers	94.9	98	0.97	0.390	0.427	0.037
66	Orlando Magic	94.8	94	1.01	0.529	0.488	0.042
67	Miami Heat	94.6	95.5	0.99	0.467	0.537	0.069
68	Portland Trail Blazers	94.1	98.4	0.96	0.349	0.390	0.041
69	Atlanta Hawks	93.7	98.4	0.95	0.336	0.366	0.030

Figure 1.4. Predicted NBA winning percentages. Exp = 13.91. See file Footballbasketballpythagoras.xls.

with them." This train of thought would lead us to believe that these teams would perform worse during the playoffs. Note that the Miami Heat and Dallas Mavericks both won about 8% more games than expected during the regular season. Therefore, we would have predicted Miami and Dallas to perform worse during the playoffs than their actual win-loss record indicated. Sure enough, both Dallas and Miami suffered unexpected first-round defeats. Conversely, during the regular season the San Antonio Spurs and Chicago Bulls won around 8% fewer games than the Pythagorean Theorem predicts, indicating that these teams would perform better than expected in the playoffs. Sure enough, the Bulls upset the Heat and gave the Detroit Pistons a tough time. Of course, the Spurs won the 2007 NBA title. In addition, the Pythagorean Theorem had the Spurs as by far the league's best team (78% predicted winning percentage). Note the team that underachieved the most was the Boston Celtics, who won nearly 9% fewer (or 7) games than predicted. Many people suggested the Celtics "tanked" games during the regular season to improve their chances of obtaining potential future superstars such as Greg Oden and Kevin Durant in the 2007 draft lottery. The fact that the Celtics won seven fewer games than expected does not prove this conjecture, but it is certainly consistent with the view that Celtics did not go all out to win every close game.

APPENDIX

Data Tables

The Excel Data Table feature enables us to see how a formula changes as the values of one or two cells in a spreadsheet are modified. This appendix shows how to use a One Way Data Table to determine how the accuracy of (2) for predicting team winning percentage depends on the value of exp. To illustrate, let's show how to use a One Way Data Table to determine how varying exp from 1 to 3 changes the average error in predicting a MLB team's winning percentage (see figure 1.2).

   Step 1. We begin by entering the possible values of exp (1, 1.1, . . . 3) in the cell range N7:N27. To enter these values, simply enter 1 in N7, 1.1 in N8, and select the cell range N8. Now drag the cross in the lower right-hand corner of N8 down to N27.

   Step 2. In cell O6 we enter the formula we want to loop through and calculate for different values of exp by entering the formula = J1.

   Step 3. In Excel 2003 or earlier, select Table from the Data Menu. In Excel 2007 select Data Table from the What If portion of the ribbon's Data tab (figure 1-a).

Figure 1-a. What If icon for Excel 2007.

   Step 4. Do not select a row input cell but select cell L2 (which contains the value of exp) as the column input cell. After selecting OK we see the results shown in figure 1.2. In effect Excel has placed the values 1, 1.1, . . . 3 into cell M2 and computed our MAD for each listed value of exp.

2

WHO HAD A BETTER YEAR, NOMAR GARCIAPARRA OR ICHIRO SUZUKI?

The Runs-Created Approach

In 2004 Seattle Mariner outfielder Ichiro Suzuki set the major league record for most hits in a season. In 1997 Boston Red Sox shortstop Nomar Garciaparra had what was considered a good (but not great) year. Their key statistics are presented in table 2.1. (For the sake of simplicity, henceforth Suzuki will be referred to as "Ichiro" or "Ichiro 2004" and Garciaparra will be referred to as "Nomar" or "Nomar 1997.")

Recall that a batter's slugging percentage is Total Bases (TB)/At Bats (AB) where

$$TB = \text{Singles} + 2 \times \text{Doubles (2B)} + 3 \times \text{Triples (3B)} + 4 \times \text{Home Runs (HR)}.$$

We see that Ichiro had a higher batting average than Nomar, but because he hit many more doubles, triples, and home runs, Nomar had a much higher slugging percentage. Ichiro walked a few more times than Nomar did. So which player had a better hitting year?

When a batter is hitting, he can cause good things (like hits or walks) to happen or cause bad things (outs) to happen. To compare hitters we must develop a metric that measures how the relative frequency of a batter's good events and bad events influence the number of runs the team scores.

In 1979 Bill James developed the first version of his famous Runs Created Formula in an attempt to compute the number of runs "created" by a hitter during the course of a season. The most easily obtained data we have available to determine how batting events influence Runs Scored are season-long team batting statistics. A sample of this data is shown in figure 2.1.

TABLE 2.1
Statistics for Ichiro Suzuki and Nomar Garciaparra

Event	Ichiro 2004	Nomar 1997
AB	704	684
Batting average	.372	.306
SLG	.455	.534
Hits	262	209
Singles	225	124
2B	24	44
3B	5	11
HR	8	30
BB+HBP	53	41

	A	B	C	D	E	F	G	H	J	S
3	Year	Runs	At Bats	Hits	Singles	2B	3B	HR	BB + HBP	Team
4	2000	864	5628	1574	995	309	34	236	655	A. Angels
5	2000	794	5549	1508	992	310	22	184	607	B. Orioles
6	2000	792	5630	1503	988	316	32	167	653	B. Red Sox
7	2000	978	5646	1615	1041	325	33	216	644	C. White Sox
8	2000	950	5683	1639	1078	310	30	221	736	C. Indians
9	2000	823	5644	1553	1028	307	41	177	605	D. Tigers
10	2000	879	5709	1644	1186	281	27	150	559	K.C. Royals
11	2000	748	5615	1516	1026	325	49	116	591	M. Twins
12	2000	871	5556	1541	1017	294	25	205	688	N.Y. Yankees
13	2000	947	5560	1501	958	281	23	239	802	O. Athletics
14	2000	907	5497	1481	957	300	26	198	823	S. Mariners
15	2000	733	5505	1414	977	253	22	162	607	T.B. Devil Rays
16	2000	848	5648	1601	1063	330	35	173	619	T. Rangers
17	2000	861	5677	1562	969	328	21	244	586	T. Blue Jays
18	2000	792	5527	1466	961	282	44	179	594	A. Diamondbacks

Figure 2.1. Team batting data for 2000 season. See file teams.xls.

James realized there should be a way to predict the runs for each team from hits, singles, 2B, 3B, HR, outs, and BB + HBP.[1] Using his great intuition, James came up with the following relatively simple formula.

---

[1] Of course, we are leaving out things like Sacrifice Hits, Sacrifice Flies, Stolen Bases and Caught Stealing. Later versions of Runs Created use these events to compute Runs Created. See http://danagonistes.blogspot.com/2004/10/brief-history-of-run-estimation-runs.html for an excellent summary of the evolution of Runs Created.

$$\text{runs created} = \frac{(\text{hits} + \text{BB} + \text{HBP}) \times (\text{TB})}{(\text{AB} + \text{BB} + \text{HBP})}. \tag{1}$$

As we will soon see, (1) does an amazingly good job of predicting how many runs a team scores in a season from hits, BB, HBP, AB, 2B, 3B, and HR. What is the rationale for (1)? To score runs you need to have runners on base, and then you need to advance them toward home plate: (Hits + Walks + HBP) is basically the number of base runners the team will have in a season. The other part of the equation, $\dfrac{\text{TB}}{(\text{AB} + \text{BB} + \text{HBP})}$, measures the rate at which runners are advanced per plate appearance. Therefore (1) is multiplying the number of base runners by the rate at which they are advanced. Using the information in figure 2.1 we can compute Runs Created for the 2000 Anaheim Angels.

$$\text{runs created} = \frac{(1{,}574 + 655) \times (995 + 2(309) + 3(34) + 4(236))}{(5{,}628 + 655)} = 943.$$

Actually, the 2000 Anaheim Angels scored 864 runs, so Runs Created overestimated the actual number of runs by around 9%. The file teams.xls calculates Runs Created for each team during the 2000–2006 seasons[2] and compares Runs Created to actual Runs Scored. We find that Runs Created was off by an average of 28 runs per team. Since the average team scored 775 runs, we find an average error of less than 4% when we try to use (1) to predict team Runs Scored. It is amazing that this simple, intuitively appealing formula does such a good job of predicting runs scored by a team. Even though more complex versions of Runs Created more accurately predict actual Runs Scored, the simplicity of (1) has caused this formula to continue to be widely used by the baseball community.

Beware Blind Extrapolation!

The problem with any version of Runs Created is that the formula is based on team statistics. A typical team has a batting average of .265, hits home runs on 3% of all plate appearances, and has a walk or HBP in around 10% of all plate appearances. Contrast these numbers to those of Barry Bonds's

[2] The data come from Sean Lahman's fabulous baseball database, http://baseball1.com/statistics/.

great 2004 season in which he had a batting average of .362, hit a HR on 7% of all plate appearances, and received a walk or HBP during approximately 39% of his plate appearances. One of the first ideas taught in business statistics class is the following: do not use a relationship that is fit to a data set to make predictions for data that are very different from the data used to fit the relationship. Following this logic, we should not expect a Runs Created Formula based on team data to accurately predict the runs created by a superstar such as Barry Bonds or by a very poor player. In chapter 4 we will remedy this problem.

Ichiro vs. Nomar

Despite this caveat, let's plunge ahead and use (1) to compare Ichiro Suzuki's 2004 season to Nomar Garciaparra's 1997 season. Let's also compare Runs Created for Barry Bonds's 2004 season to compare his statistics with those of the other two players. (See figure 2.2.)

	A	C	D	E	F	G	H	J	S	T	U
	Year	At Bats	Hits	Singles	2B	3B	HR	BB+HBP	Runs created	Game outs used	Runs created /game
225											
226	Bonds 2004	373	135	60	27	3	45	242	185.74	240.29	20.65
227	Ichiro 2004	704	262	225	24	5	8	53	133.16	451.33	7.88
228	Nomar 1997	684	209	124	44	11	30	41	125.86	500.69	6.72

Figure 2.2. Runs Created for Bonds, Suzuki, and Garciaparra. See file teams.xls.

We see that Ichiro created 133 runs and Nomar created 126 runs. Bonds created 186 runs. This indicates that Ichiro 2004 had a slightly better hitting year than Nomar 1997. Of course Bonds's performance in 2004 was vastly superior to that of the other two players.

Runs Created Per Game

A major problem with any Runs Created metric is that a bad hitter with 700 plate appearances might create more runs than a superstar with 400 plate appearances. In figure 2.3 we compare the statistics of two hypothet-

---

[3] Since the home team does not bat in the ninth inning when they are ahead and some games go into extra innings, average outs per game is not exactly 27. For the years 2001–6, average outs per game was 26.72.

	A	C	D	E	F	G	H	J	S	T	U
218	Year	At Bats	Hits	Singles	2B	3B	HR	BB + HBP	Runs created	Game outs used	Runs created/game
222	Christian	700	190	150	10	1	9	20	60.96	497.40	3.27
223	Gregory	400	120	90	15	0	15	20	60.00	272.80	5.88

Figure 2.3. Christian and Gregory's fictitious statistics.

ical players: Christian and Gregory. Christian had a batting average of .257 while Gregory had a batting average of .300. Gregory walked more often per plate appearance and had more extra-base hits. Yet Runs Created says Christian was a better player. To solve this problem we need to understand that hitters consume a scarce resource: outs. During most games a team bats for nine innings and gets 27 outs (3 outs × 9 innings = 27).[3] We can now compute Runs Created per game. To see how this works let's look at the data for Ichiro 2004 (figure 2.2).

How did we compute outs? Essentially all AB except for hits and errors result in an out. Approximately 1.8% of all AB result in errors. Therefore, we computed outs in column I as $AB - Hits - .018(AB) = .982(AB) - Hits$. Hitters also create "extra" outs through sacrifice flies (SF), sacrifice bunts (SAC), caught stealing (CS), and grounding into double plays (GIDP). In 2004 Ichiro created 22 of these extra outs. As shown in cell T219, he "used" up 451.3 outs for the Mariners. This is equivalent to $\dfrac{451.3}{26.72} = 16.9$ games. Therefore, Ichiro created $\dfrac{133.16}{16.9} = 7.88$ runs per game. More formally, runs created per game

$$= \frac{\text{runs created}}{\dfrac{.982(\text{AB}) - \text{hits} + \text{GIDP} + \text{SF} + \text{SAC} + \text{CS}}{26.72}}. \tag{2}$$

Equation (2) simply states that Runs Created per game is Runs Created by batter divided by number of games' worth of outs used by the batter. Figure 2.2 shows that Barry Bonds created an amazing 20.65 runs per game. Figure 2.2 also makes it clear that Ichiro in 2004 was a much more valuable hitter than was Nomar in 1997. After all, Ichiro created 7.88 runs per game while Nomar created 1.16 fewer runs per game (6.72 runs). We also see that Runs Created per game rates Gregory as being 2.61 runs

(5.88 − 3.27) better per game than Christian. This resolves the problem that ordinary Runs Created allowed Christian to be ranked ahead of Gregory.

Our estimate of Runs Created per game of 7.88 for Ichiro indicates that we believe a team consisting of nine Ichiros would score an average of 7.88 runs per game. Since no team consists of nine players like Ichiro, a more relevant question might be, how many runs would he create when batting with eight "average hitters"? In his book *Win Shares* (2002) Bill James came up with a more complex version of Runs Created that answers this question. I will address this question in chapters 3 and 4.

3

EVALUATING HITTERS BY LINEAR WEIGHTS

In chapter 2 we saw how knowledge of a hitter's AB, BB+HBP, singles, 2B, 3B, and HR allows us to compare hitters via the Runs Created metric. As we will see in this chapter, the Linear Weights approach can also be used to compare hitters. In business and science we often try to predict a given variable (called Y or the dependent variable) from a set of independent variables ($x_1, x_2, \ldots x_n$). Usually we try to find weights B1, B2, ... Bn and a constant that make the quantity

$$\text{Constant} + B1x_1 + B2x_2 + \ldots Bnx_n$$

a good predictor for the dependent variable.

Statisticians call the search for the weights and constant that best predict Y running a multiple linear regression. Sabermetricians (people who apply math to baseball) call the weights Linear Weights.

For our team batting data for the years 2000–2006

$$Y = \text{dependent variable} = \text{runs scored in a season.}$$

For independent variables we will use BB + HBP, singles, 2B, 3B, HR, SB [Stolen Bases]), and CS (Caught Stealing). Thus our prediction equation will look like this.

$$
\begin{aligned}
\text{predicted runs for season} = {} & \text{constant} + B1(BB + HBP) \\
& + B2(\text{singles}) + B3(2B) + B4(3B) \\
& + B5(HR) + B6(SB) + B7(CS). \quad (1)
\end{aligned}
$$

Let's see if we can use basic arithmetic to come up with a crude estimate of the value of a HR. For the years 2000–2006, an average MLB team has 38 batters come to the plate and scores 4.8 runs in a game so roughly 1 out of 8 batters scores. During a game the average MLB team has around 13 batters reach base. Therefore 4.8/13 or around 37% of all runners score. If we assume an average of one runner on base when a HR is hit, then a HR creates "runs" in the following fashion:

- The batter scores all the time instead of 1/8 of the time, which creates 7/8 of a run.
- An average of one base runner will score 100% of the time instead of 37% of the time. This creates 0.63 runs.

This leads to a crude estimate that a HR is worth around $0.87 + 0.63 = 1.5$ runs. We will soon see that our Regression model provides a similar estimate for the value of a HR.

We can use the Regression tool in Excel to search for the set of weights and constant that enable (1) to give the best forecast for Runs Scored. (See this chapter's appendix for an explanation of how to use the Regression tool.) Essentially Excel's Regression tool finds the constant and set of weights that minimize the sum over all teams of

$$(\text{actual runs scored} - \text{predicted runs scored from (1)})^2.$$

In figure 3.1, cells B17:B24 (listed under Coefficients) show that the best set of Linear Weights and constant (Intercept cell gives constant) to predict runs scored in a season is given by

$$
\begin{aligned}
\text{predicted runs} = {}& -563.03 + 0.63(\text{singles}) + 0.72(\text{2B}) \\
& + 1.24(\text{3B}) + 1.50(\text{HR}) + 0.35(\text{BB} + \text{HBP}) \\
& + 0.06(\text{SB}) + 0.02(\text{CS}). \quad\quad\quad (1)
\end{aligned}
$$

The R Square value in cell B5 indicates that the independent variables (singles, 2B, 3B, HR, BB+HBP, SB, and CS) explain 91% of the variation in the number of runs a team actually scores during a season.[1]

Equation (2) indicates that a single "creates" 0.63 runs, a double creates 0.72 runs, a triple creates 1.24 runs, a home run creates 1.50 runs, a walk or being hit by the pitch creates 0.35 runs, and a stolen base creates 0.06 runs, while being caught stealing causes 0.02 runs. We see that the HR weight agrees with our simple calculation. Also the fact that a double

---

[1] If we did not square the prediction error for each team we would find that the errors for teams that scored more runs than predicted would be canceled out by the errors for teams that scored fewer runs than predicted.

	A	B	C	D	E	F	G
1	**SUMMARY OUTPUT**						
2							
3	*Regression Statistics*						
4	Multiple R	0.954033					
5	R Square	0.910179					
6	Adjusted R Square	0.907066					
7	Standard Error	24.48612					
8	Observations	210					
9							
10	**ANOVA**						
11		*df*	*SS*	*MS*	*F*	*Significance F*	
12	Regression	7	1227267	175323.912	292.4162	4.9885E−102	
13	Residual	202	121113.1	599.569857			
14	Total	209	1348380				
15							
16		Coefficients	Standard Error	t Stat	P-value	Lower 95%	Upper 95%
17	Inter-ceptions	−563.029	37.21595	−15.128695	4.52E−35	−636.4104075	−489.647257
18	Singles	0.625452	0.031354	19.9479691	1.23E−49	0.563628474	0.687275336
19	Doubles	0.720178	0.069181	10.4099998	1.36E−20	0.583767923	0.856588501
20	Triples	1.235803	0.203831	6.06288716	6.47E−09	0.833894343	1.637712396
21	Home Runs	1.495572	0.061438	24.3426548	5.48E−62	1.374428861	1.616714188
22	Walks + Hit by Pitcher	0.346469	0.025734	13.4633465	6.55E−30	0.295726467	0.397210735
23	Stolen Bases	0.05881	0.07493	0.78485776	0.433456	−0.088936408	0.206555885
24	Caught Stealing	0.015257	0.189734	0.08040989	0.935991	−0.358857643	0.389370703

Figure 3.1. Regression output with CS and SB included. The results of the regression are in sheet Nouts of workbook teamsnocssbouts.xls.

is worth more than a single but less than two singles is reasonable. The fact that a single is worth more than a walk makes sense because singles often advance runners two bases. It is also reasonable that a triple is worth more than a double but less than a home run. Of course, the positive coefficient for CS is unreasonable because it indicates that each time a base runner is caught stealing he creates runs. This anomaly will be explained shortly.

The Meaning of P-Values

When we run a regression, we should always check whether or not each independent variable has a significant effect on the dependent variable. We do this by looking at each independent variable's p-value. These are shown in column E of figure 3.1. Each independent variable has a p-value between 0 and 1. Any independent variable with a p-value $< .05$ is considered a useful predictor of the dependent variable (after adjusting for the other independent variables). Essentially the p-value for an independent variable gives the probability that (in the presence of all other independent variables used to fit the regression) the independent variable does not enhance our predictive ability. For example, there is only around one chance in 10^{20} that doubles do not enhance our ability for predicting Runs Scored even after we know singles, 3B, HR, BB+HBP, CS, and SB. Figure 3.1 shows that all independent variables except for SB and CS have p-values that are very close to 0. For example, singles have a p-value of 1.23×10^{-49}. This means that singles almost surely help predict team runs even after adjusting for all other independent variables. There is a 43% chance, however, that SB is not needed to predict Runs Scored and an almost 94% chance that CS is not needed to predict Runs Scored. The high p-values for these independent variables indicate that we should drop them from the regression and rerun the analysis. For example, this means that the surprisingly positive coefficient of .02 for CS in our equation was just a random fluctuation from a coefficient of 0. The resulting regression is shown in figure 3.2.

All of the independent variables have p-values $< .05$, so they all pass the test of statistical significance. Let's use the following equation (derived from cells B17:B22 of figure 3.2) to predict runs scored by a team in a season.

$$\text{predicted runs for a season} = -560 + .63(\text{singles})$$
$$+ 0.71(2B) + 1.26(3B)$$
$$+ 1.49(HR) + 0.35(BB + HBP).$$

Note our R Square is still 91%, even after dropping CS and SB as independent variables. This is unsurprising because the high p-values for these independent variables indicated that they would not help predict Runs Scored after we knew the other independent variables. Also note that our HR weight of 1.49 almost exactly agrees with our crude estimate of 1.5.

	A	B	C	D	E	F	G
1	**SUMMARY OUTPUT**						
2							
3	*Regression Statistics*						
4	Multiple R	0.953776					
5	R Square	0.909688					
6	Adjusted R Square	0.907475					
7	Standard Error	24.48223					
8	Observations	210					
9							
10	**ANOVA**						
11		*df*	*SS*	*MS*	*F*	*Significance F*	
12	Regression	5	1226606	245321.1319	410.9687	2.0992E–104	
13	Residual	204	121774.5	596.9340126			
14	Total	209	1348380				
15							
16		Coefficients	Standard Error	t Stat	P-value	Lower 95%	Upper 95%
17	Inter-ceptions	–559.997	35.52184	–15.76486473	3.81E–37	–630.0341104	–489.9600492
18	Singles	0.632786	0.030209	20.94664121	9.77E–53	0.573222833	0.692348228
19	Doubles	0.705947	0.067574	10.44707819	9.74E–21	0.572714992	0.839179681
20	Triples	1.263721	0.200532	6.301838725	1.78E–09	0.868340029	1.65910294
21	Home Runs	1.490741	0.060848	24.49945673	1.1E–62	1.370769861	1.610712843
22	Walks + Hit by Pitcher	0.346563	0.025509	13.58610506	2.3E–30	0.296268954	0.396857822

Figure 3.2. P-values for Linear Weights regression. See sheet Noutscssb of workbook teamsnocssbouts.xls.

Accuracy of Linear Weights vs. Runs Created

Do Linear Weights do a better job of forecasting Runs Scored than does Bill James's original Runs Created Formula? We see in cell D2 of figure 3.3 that for the team hitting data (years 2000–2006) Linear Weights was off by an average of 18.63 runs (an average of 2% per team) while, as previously noted, Runs Created was off by 28 runs per game. Thus, Linear Weights do a better job of predicting team runs than does basic Runs Created.

	A	B	C	D	E	F	G	H	I	J	K	L
1				**MAD**			**Linear Weights**					
2				18.63392992			0.632785531	0.705947	1.2637	1.49074135	0.346563388	
3	Year	Runs	Predicted Runs	Absolute Error	At Bats	Hits	Singles	2B	3B	HR	BB + HBP	Team
4	2000	864	909.5427592	45.54275916	5628	1574	995	309	34	236	655	A. Angels
5	2000	794	799.0320991	5.032099146	5549	1508	992	310	22	184	607	B. Orioles
6	2000	792	803.9731688	11.97316875	5630	1503	988	316	32	167	653	B. Red Sox
7	2000	978	915.0553052	62.94469483	5646	1615	1041	325	33	216	644	C. White Sox
8	2000	950	963.4255338	13.42553378	5683	1639	1078	310	30	221	736	C. Indians
9	2000	823	832.5769282	9.576928216	5644	1553	1028	307	41	177	605	D. Tigers
10	2000	879	840.3183782	38.6816218	5709	1644	1186	281	27	150	559	K.C. Royals
11	2000	748	758.3410712	10.34107117	5615	1516	1026	325	49	116	591	M. Twins
12	2000	871	866.7249473	4.275052678	5556	1541	1017	294	25	205	688	N.Y. Yankees
13	2000	947	907.8792749	39.12072513	5560	1501	958	281	23	239	802	O. Athletics
14	2000	907	870.6080889	36.3919111	5497	1481	957	300	26	198	823	S. Mariners
15	2000	733	716.5050083	16.49499174	5505	1414	977	253	22	162	607	T.B. Devil Rays
16	2000	848	862.2678037	14.26780365	5648	1601	1063	330	35	173	619	T. Rangers
17	2000	861	878.0880125	17.08801249	5677	1562	969	328	21	244	586	T. Blue Jays
18	2000	792	775.4920641	16.50793587	5527	1466	961	282	44	179	594	A. Diamondbacks

Figure 3.3. Measuring accuracy of Linear Weights. See sheet accuracy Linear Weights of file teamsnocssbouts.xls.

The History of Linear Weights

Let's briefly trace the history of Linear Weights. In 1916 F. C. Lane, the editor of *Baseball Magazine*, used the records of how 1,000 hits resulted in advancing runners around the bases to come up with an estimate of Linear Weights. During the late 1950s and 1960s, military officer George Lindsay looked at a large set of game data and came up with a set of Linear Weights. Then in 1978 statistician Pete Palmer used a Monte Carlo simulation model (see chapter 4) to estimate the value of each type of baseball event. During 1989 *Washington Post* reporter Thomas Boswell also came up with a set of Linear Weights.[2] The weights obtained by these pioneers are summarized in table 3.1.

For reasons I will discuss in chapter 4, I believe Monte Carlo simulation (as implemented by Palmer) is the best way to determine Linear Weights.

[2] See Dan Agonistes's excellent summary, http://danagonistes.blogspot.com/2004/10/brief-history-of-run-estimation.html; Schwarz, *The Numbers Game*; Palmer, *The Hidden Game of Baseball*; and Boswell, *Total Baseball*.

TABLE 3.1
The Historical Evolution of Linear Weights Estimates

Event	Lane	Lindsay	Palmer	Boswell	Our Regression
BB+HBP	0.164	—	0.33	1.0	0.35
Singles	0.457	0.41	0.46	1.0	0.63
2B	0.786	0.82	0.8	2.0	0.71
3B	1.15	1.06	1.02	3.0	1.26
HR	1.55	1.42	1.4	4.0	1.49
Outs	—	—	−0.25	−1.0	—
SB	—	—	0.3	1.0	—
CS	—	—	−0.6	−1.0	—

Note: Empty fields indicate events the authors have not used in a specific model.

	A	B	C	D	E	F	G	H	J	S	T
233	**Year**	**Scale Factor**	**At Bats**	**Hits**	**Singles**	**2B**	**3B**	**HR**	**BB + HBP**	**Linear Weights Run**	**Runs per game**
234	Bonds 2004	18.016031	6719.98	2432	1081	486.4	54	810.72139	4359.879477	3259.26522574	20.11892114
235	Ichiro 2004	9.5916938	6752.55	2513	2158	230.2	48	76.733551	508.3597738	1323.318592	8.168633281
236	Nomar 1997	8.646103	5913.93	1807	1072	380.4	95.1	259.38309	354.4902215	1020.697841	6.300603957

Figure 3.4. Linear Weights estimates of runs per game created by Bonds, Suzuki, and Garciaparra.

Despite this, let's use our regression to evaluate hitters. Recall that (2) predicted runs scored given a team's statistics for an entire season. How can we use (2) to predict how many runs could be scored if a team consisted entirely of, say Barry Bonds (2004), Ichiro Suzuki (2004), or Nomar Garciaparra (1997)? Let's look at Bonds 2004 first (figure 3.4).

Bonds 2004 made 240.29 outs. As explained in chapter 2, we computed outs made by a hitter as .982(AB) + SF + SAC + CS + GIDP. Given an average of 26.72 outs per game, a team's season has $26.72 \times 162 = 4{,}329$ outs. Bonds hit 45 HR. So for each out he hit 45/240.29 = .187 HR. Thus for a whole season we would predict a team of nine Barry Bonds to hit $4{,}329 \times (45/240.29) = 811$ HR. Now we see how to use (2) to predict runs scored by a team consisting entirely of that player.[3] Simply "scale up" each of Bonds's statistics by the following:

$$4{,}329/240.29 = 18.02 = \text{outs for season/player outs.}$$

In rows 233–35 each player's statistics (from rows 226–28) were multiplied by $4{,}329/(\text{player's outs})$. This is a player's "scale factor." Then in column S the Linear Weights regression model (equation 2) was applied to the data in rows 233–35 to predict total season runs for a team consisting of the single player (see cells S233:S235). In cells T233:T235 the predicted runs for a season are divided by 162 to create a predicted runs per game. We predict a team of Bonds 2004 to score 20.12 runs per game, a team of Ichiro 2004 to score 8.17 runs per game, and a team of Nomar 1997 to score 6.30 runs per game. Note that using Runs Created gives estimates of 20.65, 7.88, and 6.72 runs, respectively, for the three players. Thus for the three players Runs Created and Linear Weights give very similar predictions for the number of runs a player is responsible for during a game.

OBP, SLG, OBP + SLG, and Runs Created

As Michael Lewis brilliantly explains in his best-seller *Moneyball*, during the 1980s and 1990s MLB front offices came to realize the importance of On-Base Percentage (OBP) as a measure of a hitter's effectiveness. OBP is simply the fraction of a player's plate appearances in which he reaches base on a hit, walk, or HBP. During the 2000–2006 seasons the average OBP was 0.33. OBP is a better measure of hitting effectiveness than ordinary batting average because a player with a high OBP uses less of a team's scarce resource (outs). Unfortunately, many players with a high OBP (such as Ty Cobb and Willie Keeler) do not hit many home runs, so their value is overstated by simply relying on OBP. Therefore, baseball experts created a new statistic: On-Base Plus Slugging (OPS), which is slugging percentage, or SLG (TB/AB), plus OBP. The rationale is that by including SLG in OPS we give proper credit to power hitters. In 2004 OPS "arrived" when it was included on Topps baseball cards.

---

[3] It might be helpful to note that

$$\left(\frac{\text{player HRs}}{\text{season}}\right) = \left(\frac{\text{player HRs}}{\text{player outs}}\right)\left(\frac{\text{total outs}}{\text{season}}\right).$$

Note that both sides of this equation have the same units.

	A	B	C	D	E	F	G
1	**SUMMARY OUTPUT**						
2							
3	*Regression Statistics*						
4	**Multiple R**	0.9520351					
5	**R Square**	0.9063709					
6	**Adjusted R Square**	0.9053129					
7	**Standard Error**	25.70605					
8	**Observations**	180					
9							
10	**ANOVA**						
11		*df*	*SS*	*MS*	*F*	*Significance F*	
12	**Regression**	2	1132241	566120.6	856.7187	9.32975E–92	
13	**Residual**	177	116961.8	660.801			
14	**Total**	179	1249203				
15							
16		*Coefficients*	*Standard Error*	*t Stat*	*P-value*	*Lower 95%*	*Upper 95%*
17	**Intercept**	–1003.647	49.63353	–20.2211	7.05E–48	–1101.596424	–905.6971482
18	**Slugging %**	1700.8005	121.8842	13.95424	2.49E–30	1460.267357	1941.333699
19	**On Base %**	3156.7146	232.9325	13.55206	3.67E–29	2697.032329	3616.39681

Figure 3.5. Regression predicting team runs from OBP and SLG. See file teamhittingobsslug.xls.

Of course, OPS gives equal weight to SLG and OBP. Is this reasonable? To determine the proper relative weight to give SLG and OBP I used 2000–2006 team data and ran a regression to predict team Runs Scored using OBP and SLG as independent variables.

Figure 3.5 shows that both OBP and SLG are highly significant (each has a p-value near 0). The R Square in cell B5 indicates that we explain 90.6% of the variation in Runs Scored. This compares very favorably with the best Linear Weights model, which had an R Square of .91. Since this model seems easier to understand, it is easy to see why OBP and SLG are highly valued by baseball front offices. Note, however, that we predict team Runs Scored as $-1003.65 + 1700.8 \times (\text{SLG}) + 3{,}157 \times (\text{OBP})$. This indicates that OBP is roughly twice as important ($3{,}157/1{,}700$ is near 2) as SLG. Perhaps the baseball cards should include a new statistic: $2 \times \text{OBP} + \text{SLG}$.

Runs Created above Average

One way to evaluate a player such as Ichiro 2004 is to ask how many more runs an average MLB team would score if Ichiro 2004 were added to the team (see figure 3.6). After entering a player's batting statistics in row 7, cell E11 computes the number of runs the player would add to an average MLB team. Let's examine the logic underlying this spreadsheet.

Row 7 shows the number of singles, 2B, 3B, HR, BB + HBP and total outs made by Ichiro 2004. We see that Ichiro created 451 outs. Row 6 shows the same statistics for an average MLB team (based on 2000–2006 seasons).

If we add Ichiro to an average team, the rest of the "average players" will create $4328.64 - 451 = 3877.64$ outs. Let $3877.64/4328.64 = .896$ be defined as teammult. Then the non-Ichiro plate appearances by the remaining members of our average player plus the Ichiro 2004 team will create teammult $\times 972.08$ singles, teammult $\times 296$ doubles, and so forth. Thus, our Ichiro 2004 + average player team will create $225 +$ teammult $\times 972.08 = 1095.7$ singles, $24 +$ teammult $\times 296 = 289.13$ doubles, and so forth. This implies that our Ichiro 2004 + average player team is predicted by Linear Weights to score the following number of runs.

	A	C	D	E	F	G	H	I	J	K
1						Outs Used				
2						451.328				
4	teammult		Intercept	Singles	2B	3B	HR	BB + HBP		
5	0.8957345	Linear Weights	−556	0.63279	0.7059	1.2637215	1.49074135	0.34656339	Outs	Predicted Runs Scored
6		Average Team		972.081	296	30.82381	177.480952	599.87619	4329	779.5018417
7		Ichiro		225	24	5	8	60	451	
8		Ichiro Added to Average Team		1095.73	289.13	32.609948	166.975805	597.329774		0.034
9										
10				Ichiro Runs Over Average						
11				59.111						

Figure 3.6. Computing how many runs Ichiro would add to an average team. See file Ichiroaboveaverage.xls.

$$-556 + .633 \times (1095.7) + (.706) \times 289.13$$
$$+ (1.264) \times 32.61 + (1.491) \times (166.98)$$
$$+ (.3466) \times 597.33 = 838.61.$$

Since an average team was predicted by Linear Weights to score 779.50 runs, the addition of Ichiro 2004 to an average team would add 838.61 − 779.50 = 59.11 runs. Thus we estimate that adding Ichiro 2004 to an average team would add around 59 runs. This estimate of Ichiro's hitting ability puts his contribution into the context of a typical MLB team, and therefore seems more useful than an estimate of how many runs would be scored by a team made up entirely of Ichiro 2004.

Figure 3.7 lists the top twenty-five Runs above Average performances (for players with at least 350 AB) during the 2001–6 seasons. Note the incredible dominance of Barry Bonds; he had the top four performances. Albert Pujols had four of the top twenty-five performances while Todd Helton had three of the top twelve performances.

	A	B	C	D
2	Rank	Year	Player	Runs above average
3	1	2004	B. Bonds	178.72
4	2	2002	B. Bonds	153.8278451
5	3	2001	B. Bonds	142.2021593
6	4	2003	B. Bonds	120.84
7	5	2001	S. Sosa	112.4092099
8	6	2001	L. Gonzalez	99.30956815
9	7	2006	R. Howard	96.70402992
10	8	2001	J. Giambi	96.64777824
11	9	2003	T. Helton	92.16893785
12	10	2004	T. Helton	91.33935918
13	11	2003	A. Pujols	90.72817498
14	12	2001	T. Helton	87.85495932
15	13	2002	J. Thome	85.33958204
16	14	2006	A. Pujols	84.69690329
17	15	2005	D. Lee	84.5746433
18	16	2000	B. Bonds	83.66
19	17	2005	A. Pujols	82.23517954
20	18	2001	L. Walker	78.65841316
21	19	2002	B. Giles	78.07581834
22	20	2005	A. Rodriguez	77.69034834
23	21	2006	D. Ortiz	76.44267022
24	22	2003	C. Delgado	75.87692757
25	23	2001	C. Jones	75.55723654
26	24	2004	A. Pujols	74.32012661
27	25	2003	M. Ramirez	74.04277236

Figure 3.7. The Top Runs above Average Performances, 2001−6.

In chapter 4 we will use Monte Carlo simulation to obtain another estimate of how many runs a player adds to a particular team.

APPENDIX

Running Regressions in Excel

To run regressions in Excel it is helpful to install the Analysis Toolpak Add-In.

Installation of the Analysis Toolpak

To install the Analysis Toolpak Add-In in Excel 2003 or an earlier version of Excel, select Add-Ins from the Tools menu and check the Analysis Toolpak option. Checking OK completes the installation.

To install the Analysis Toolpak in Excel 2007 first select the Office button (the oval button in the left-hand corner of the ribbon). Then choose Excel Options followed by Add-Ins. Now hit Go and check Analysis Toolpak and choose OK.

Figure 3-a. Office button.

Running a Regression

The regression shown in figure 3.1 predicts team Runs Scored from a team's singles, 2B, 3B, HR, BB + HBP, SB, and CS. To run the regression, first go to the sheet Team of the workbook teamsnocsouts.xls. In Excel 2003 or earlier bring up the Analysis Toolpak by choosing Tools and then Data Analysis. In Excel 2007 bring up the Analysis Toolpak by selecting the Data Tab and then Choosing Data Analysis from the right-hand portion of the tab.

Now select the regression option and fill in dialog box as shown in figure 3-b. This tells Excel we want to predict the team Runs Scored (in cell

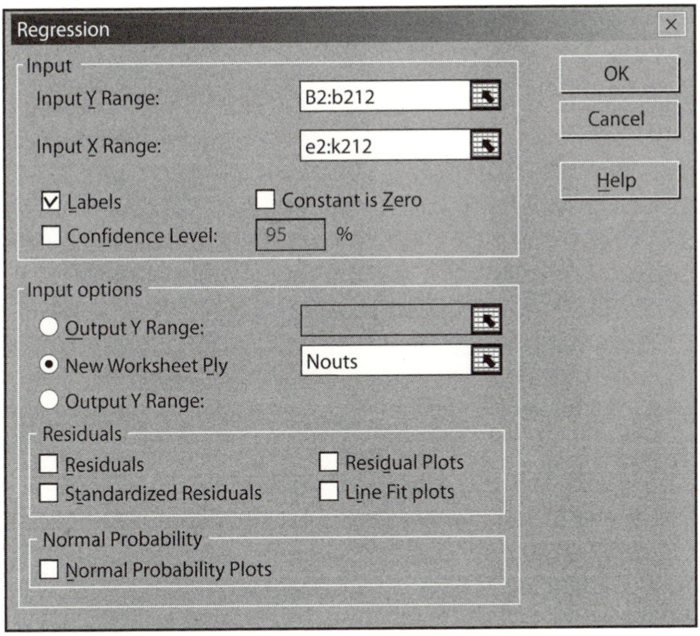

Figure 3-b. Regression dialog box.

range B3:B212) using the independent variables in cell range E3:K212 (singles, 2B, 3B, HR, BB + HBP, SB, and CS). We checked the Labels box so that our column labels shown in row 2 will be included in the regression output. The output (as shown in figure 3.1) will be placed in the worksheet Nouts.

4

EVALUATING HITTERS BY MONTE CARLO SIMULATION

In chapters 2 and 3 we showed how to use Runs Created and Linear Weights to evaluate a hitter's effectiveness. These metrics were primarily developed to "fit" the relationship between runs scored by a team during a season and team statistics such as walks, singles, doubles, triples, and home runs. We pointed out that for players whose event frequencies differ greatly from typical team frequencies, these metrics might do a poor job of evaluating a hitter's effectiveness.

A simple example will show how Runs Created and Linear Weights can be very inaccurate.[1] Consider a player (let's call him Joe Hardy after the hero of the wonderful movie and play *Damn Yankees*) who hits a home run at 50% of his plate appearances and makes an out at the other 50% of his plate appearances. Since Joe hits as many home runs as he makes outs, you would expect Joe "on average" to alternate HR, OUT, HR, OUT, HR, OUT, for an average of 3 runs per inning. In the appendix to chapter 6 we will use the principle of conditional expectation to give a mathematical proof of this result.

In 162 nine-inning games Joe Hardy will make, on average, 4,374 outs ($162 \times 27 = 4{,}374$) and hit 4,374 home runs. As shown in figure 4.1, we find that Runs Created predicts that Joe Hardy would generate 54 runs per game (or 6 per inning) and Linear Weights predicts Joe Hardy to generate 36.77 runs per game (or 4.08 runs per inning). Both estimates are far from the true value of 27 runs per game.

Introduction to Monte Carlo Simulation

How can we show that our player generates 3 runs per inning, or 27 runs per game? We can do so by programming the computer to play out many

---

[1] This was described to me by Jeff Sagarin, *USA Today* sports statistician.

	K	L	M	N	O	P
3	**Method**	**At Bats**	**HR**	**Outs**	**Runs Created**	**Runs Created / Game**
4	Bill James	8748	4374	4374	8748	54
5	Linear Weights	8748	4374	4374	5957.26	36.77321

Figure 4.1. Runs Created and Linear Weights predicted runs per game for Joe Hardy. See file simulationmotivator.xls.

innings and averaging the number of runs scored per inning. Developing a computer model to repeatedly play out an uncertain situation is called Monte Carlo simulation.

Physicists and astronomers use this model to simulate the evolution of the universe. Biologists use the model to simulate the evolution of life on earth. Corporate financial analysts use Monte Carlo simulation to evaluate the likelihood that a new GM vehicle or a new Proctor & Gamble shampoo will be profitable. Wall Street rocket scientists use Monte Carlo simulation to price exotic or complex financial derivatives. The term "Monte Carlo simulation" was coined by the Polish-born physicist Stanislaw Ulam, who used Monte Carlo simulation in the 1930s to determine the chance of success for the chain reaction needed for an atom bomb to detonate successfully. Ulam's simulation was given the military code name Monte Carlo, and the name Monte Carlo simulation has been used ever since.

How can we play out an inning? Simply flip a coin and assign a toss of heads to an out and a toss of tails to a home run. Or we could draw from a deck of cards and assign a red card to an out and a black card to a home run. Both the coin toss and the card-drawing method will assign a 0.5 chance to a home run and a 0.5 chance to an out. We keep flipping the coin or drawing a card (with replacement) until we obtain 3 outs. Then we record the number of home runs. We repeat this procedure about 1,000 times and average the number of runs scored per inning. This average should closely approximate the average runs per inning scored by our hypothetical player. We will get very close to 3,000 total runs, which yields an estimate of 3 runs per inning. I implemented the simple Monte Carlo simulation using Microsoft Excel. (See figure 4.2.) Excel contains a function RAND(). If you type = RAND() in any cell and hit the F9 key, the number in the cell will change. The RAND() function yields any number between 0 and 1 with equal probability. This means, for example, that half the time RAND() yields a number between 0 and 0.5, and

	B	C	D	E	F	G	H
2	Batter	Random Number	Result	Outs	Runs	Over?	Total Runs
3	1	0.31683256	HR	0	1	no	2
4	2	0.51244762	out	1	1	no	
5	3	0.45037806	HR	1	2	no	
6	4	0.634642925	out	2	2	no	
7	5	0.785525468	out	3	2	yes	

Figure 4.2. Simulating one inning for Joe Hardy. See file simulationmotivator.xls.

half the time RAND() yields a number between 0.5 and 1. The results generated by the RAND() function are called random numbers. Therefore, we can simulate an inning for our player by assigning an outcome of a home run to a random number less than or equal to 0.5 and assigning an outcome of an out to a random number between 0.5 and 1. By hitting F9 in spreadsheet simulationmotivator.xls, you can see the results of a simulated inning (see figure 4.2). For our simulated inning, each random number less than or equal to 0.5 yielded a home run and any other random number yielded an out. For our simulated inning, 2 runs were scored.

Cells J6:J1005 contain the results of 1,000 simulated innings, while cell J3 contains the average runs per inning generated during our 1,000 hypothetical innings. The chapter appendix explains how Excel's Data Table feature was used to perform the simulation 1,000 times. Whenever you hit F9, you will see cell J3 is very close to 3, indicating that our player will generate around 3 runs per inning, or 27 runs per game (not 54 runs per game as Runs Created predicts).

Simulating Runs Scored by a Team of Nine Ichiros

Buoyed by the success of our simple simulation model, we can now simulate the number of runs that would be scored by a team of, say, nine Ichiro 2004s. We need to follow through the progress of an inning and track the runners on base, runs scored, and number of outs. In our model the events that can occur at each plate appearance are displayed in figure 4.3.

- We assume each error advances all base runners a single base.
- A long single advances each runner two bases.
- A medium single scores a runner from second base but advances a runner on first only one base.
- A short single advances all runners one base.

	C	D
13		**Event**
14	1	Strikeout
15	2	Walk
16	3	Hit by pitch
17	4	Error
18	5	Long single (advance 2 bases)
19	6	Medium single (score from 2nd)
20	7	Short single (advance one base)
21	8	Short double
22	9	Long double
23	10	Triple
24	11	Home run
25	12	Ground into double play
26	13	Normal ground ball
27	14	Line drive or infield fly
28	15	Long fly
29	16	Medium fly
30	17	Short fly

Figure 4.3. Event codes for baseball simulations. See file Ichiro04may28.xls.

- A short double advances each runner two bases.
- A long double scores a runner from first.
- GIDP is a ground ball double play if there is a runner on first, first and second, or first and third, or if the bases are loaded. In other situations the batter is out and the other runners stay where they are.
- Normal GO is a ground out that results in a force out with a runner on first, first and second, or first and third, or if the bases are loaded. We assume that with runners on second and third the runners stay put; with a runner on third the runner scores; and with a runner on second the runner advances to third.
- A long fly ball advances (if there are fewer than two outs) a runner on second or third one base.
- A medium fly ball (if there are fewer than two outs) scores a runner from third.
- A short fly or line drive infield fly does not advance any runners.

Next we need to assign probabilities to each of these events. During recent seasons approximately 1.8% of all AB have resulted in an error. Each player's information is input in cells E3 and E6:E12. Let's input Ichiro's 2004 statistics. (See figure 4.4.) For Ichiro, AB + SB + SF = 704 + 2 + 3 = 709. He walked 49 times, hit 225 singles, etc.

	D	E	F
1		Number	Probability
2	Plate Appearances	762	
3	At Bats +Sac. Hits + Sac. Bunts	709	
4	Errors	13	0.0170604
5	Outs (in play)	371	0.4868766
6	Strikeouts	63	0.0826772
7	BB	49	0.0643045
8	HBP	4	0.0052493
9	Singles	225	0.2952756
10	2B	24	0.0314961
11	3B	5	0.0065617
12	HR	8	0.0104987

Figure 4.4. Inputs to Ichiro simulation. The simulation omits relatively infrequent baseball events such as steals, caught stealing, passed balls, wild pitches, balks, and so forth.

Outs (in play) are plate appearances that result in non-strikeout outs: outs (in play) = (AB + SF + SB) − hits − errors − strikeouts.

Historically, errors are 1.8% of AB + SB + SF, so we compute errors = .018 × (AB + SB + SF).

Also, total plate appearances = BB + HBP + (AB + SB + SF), or 709 + 49 + 4 = 762.

We can now compute the probability of various events as (frequency of event)/(total plate appearances). For example, we estimate the probability of an Ichiro single as 225/762 = .295.

We also need to estimate probabilities for all possible types of singles, doubles, and outs in play. For example, what fraction of outs in play are GIDP? Using data from Earnshaw Cook's *Percentage Baseball* (1966) and discussions with Jeff Sagarin (who has built many accurate baseball simulation models), we estimated these fractions as follows:

- 30% of singles are long singles, 50% are medium singles, and 20% are short singles.
- 80% of doubles are short doubles and 20% are long doubles.
- 53.8% of outs in play are ground balls, 15.3% are infield flies or line drives, and 30.9% are fly balls.
- 50% of ground outs are GIDP and 50% are Normal GOs.
- 20% of all fly balls are long fly balls, 50% are medium fly balls, and 30% are short fly balls.

To verify that these parameters are accurate, I simulated 50,000 innings using the composite MLB statistics for the 2006 season from the teams.xls file. The results showed that simulated runs per game were within 1% of the actual runs per game.

Let's use the Excel simulation add-in @RISK to "play out" an inning thousands (or millions) of times. Basically @Risk generates the event for each plate appearance based on the probabilities that are input (of course, these probabilities are based on the player we wish to evaluate). For each plate appearance, @RISK generates a random number between 0 and 1. For example, for Ichiro a random number less than or equal to 0.295 would yield a single. This will cause 29.5% of Ichiro's plate appearances (as happened during the actual 2004 season) to result in a single. In a similar fashion the other possible batter outcomes will occur in the simulation with the same probability as they actually occurred.

Two sample innings of our Ichiro 2004 simulation are shown in figure 4.5. The Entering State column tracks the runners on base; for example, 101 means runner on first and third while 100 means runner on first. The Outcome column tracks the outcome of each plate appearance using the codes shown in figure 4.3. For example, event code 6 represents a medium single.

In the first inning shown in figure 4.5, our team of nine Ichiros scored three runs. In the second inning shown, the team scored no runs. Playing out thousands of innings with @RISK enables us to estimate the average number of runs scored per inning by a team of nine Ichiros. Then we multiply the average number of innings a team bats during a game (26.72/3) to estimate the number of Runs Created per game by Ichiro. Since we are playing out each inning using the actual probabilities corresponding to a given player, our Monte Carlo estimate of the runs per inning produced by nine Ichiros (or nine of any other player) should be a far better estimate than Runs Created or Linear Weights. The Monte Carlo estimate of runs per game should be accurate for *any player, no matter how good or bad*. As we have shown with our Joe Hardy example, the accuracy of Runs Created and Linear Weights as measures of hitting effectiveness breaks down for extreme cases.

Simulation Results for Ichiro, Nomar, and Bonds

For Ichiro 2004, Nomar 1997, and Bonds 2004 our simulation yields the following estimates for Runs Created per game.

- Ichiro 2004: 6.92 runs per game
- Nomar 1997: 5.91 runs per game
- Bonds 2004: 21.02 runs per game

(a)

	B	C	D	E	F	G	H	I	J
54							end	7	
55							runs	3	
56	Batter	Entering state	Outcome	State #	Outcome #	Runs	Outs made	Outs	Done?
57	1	000	Short Double	1	8	0	0	0	no
58	2	010	Long Single (advanced 2 bases)	6	5	1	0	0	no
59	3	100	Medium Single (score from 2nd)	2	6	0	0	0	no
60	4	110	Medium Fly	3	16	0	1	1	no
61	5	110	Short Double	3	8	1	0	1	no
62	6	011	Error	5	4	1	0	1	no
63	7	101	Ground into Double Play	4	12	0	2	3	yes

(b)

	B	C	D	E	F	G	H	I	J
54							end	5	
55							runs	0	
56	Batter	Entering state	Outcome	State #	Outcome #	Runs	Outs made	Outs	Done?
57	1	000	Error	1	4	0	0	0	no
58	2	100	Long Fly	2	15	0	1	1	no
59	3	100	Short Fly	2	17	0	1	2	no
60	4	100	Medium Single (score from 2nd)	2	6	0	0	2	no
61	5	110	Strikeout	3	1	0	1	3	yes

Figure 4.5. Two sample innings of Ichiro 2004 simulation.

There is a problem with our Bonds 2004 result, however. Barry Bonds received 232 walks during 2004. However, 120 of those were intentional because pitchers would rather pitch to the other players, who were not as good at hitting as Bonds. For a team consisting of nine Bonds 2004s there would be no point in issuing an intentional walk. Therefore, we reran our simulation after eliminating the intentional walks from Bonds's statistics and found that Bonds created 15.98 runs per game.

How Many Runs Did Albert Pujols Add to the St. Louis Cardinals in 2006?

Of course, there will never be team of nine Ichiros, nine Bonds, or nine Nomars. What we really want to know is how many runs a player adds to his team. Let's try and determine how many runs Albert Pujols added to the 2006 St. Louis Cardinals (let's call him Pujols 2006). The hitting

	H	I
2	**Outcome**	**Number**
3	Plate Appearances	5591
4	At Bats + Sac. Hits + Sac. Bunts	5095
5	Errors	92
6	Outs (in Play)	2824
7	Strikeouts	872
8	BB	439
9	HPB	57
10	Singles	887
11	2B	259
12	3B	26
13	HR	135

Figure 4.6. St. Louis Cardinal statistics (without Pujols), 2006. See file Pujolsmay26.xls.

	B	E	D
2	**Outcome**	**Number**	**Probability**
3	Plate Appearances	634	
4	At Bats +Sac. Hits + Sac. Bunts	538	
5	Errors	10	0.015773
6	Outs (in play)	301	0.474763
7	Strikeouts	50	0.078864
8	BB	92	0.14511
9	HBP	4	0.006309
10	Singles	94	0.148265
11	2B	33	0.05205
12	3B	1	0.001577
13	HR	49	0.077287

Figure 4.7. Albert Pujols's 2006 statistics. See file Pujolsmay26.xls.

statistics for the 2006 Cardinals (excluding Pujols) are shown in figure 4.6 and those for Pujols are shown in figure 4.7.

Using both figures, note that 7.7% of Pujols's plate appearances resulted in home runs, but for the 2006 Cardinals without Pujols, only 2.4% of all plate appearances resulted in home runs. We can now estimate how many runs Pujols added to the St. Louis Cardinals. Without Pujols we assume that each hitter's probabilities are governed by the data in figure 4.6. Playing out 25,000 innings (based on the runs per inning from our simulation), the Cardinals were projected to score an average of 706 runs without Pujols. With Pujols, the Cardinals actually scored 781 runs. How many wins can we estimate that Pujols added, compared to what an average Cardinal hitter adds? Let's use the Pythagorean Theorem from chapter 1. During 2006 the Cardinals gave up 762 runs. This yields a scoring ratio of $781/762 = 1.025$. Since the Cardinals played only 161 games during 2006, the Pythagorean Theorem predicts they should have won $\dfrac{161 \times 1.025^2}{1.025^2 + 1} = 82.48$ games.

Without Pujols our simulation yielded a scoring ratio of $706/762 = .927$. Therefore, with Pujols the Pythagorean Theorem predicts that the Cardinals would have won $\dfrac{161 \times .927^2}{.927^2 + 1} = 74.36$ games. Thus our model estimates that Pujols added $82.48 - 74.36 = 8.12$ wins for the Cardinals (assuming that Pujols's plate appearances were replaced by an average non-Pujols Cardinal hitter).

Pujols vs. the Average Major Leaguer

In his *Historical Baseball Abstract* Bill James advocates comparing a player to an "average major leaguer." Let's try to determine how many extra runs an "average 2006" team would score if we replaced 634 of the average team's plate appearances with Pujols's statistics (shown in figure 4.7). The file Pujolsoveraverage.xls allows us to input two sets of player statistics. We input Pujols's 2006 statistics in cells B2:B12. Then we input the average 2006 MLB team's statistics in H2:I12. See figure 4.8.

We can see that Pujols hit many more home runs, had many more walks, and had fewer strikeouts per plate appearance than the average 2006 batter. When simulating an inning, each batter's probabilities will be generated using either the player data from column D or the team data in column J. Since Pujols had 634 plate appearances and the average team had 6,236 plate appearances, we choose each batter to be Pujols (column D data) with probability $634/6,236 = .102$, and choose each batter to be an "average batter" (column J data) with probability $1 - .102 = .898$. After running 50,000 innings for the average team and the team replacing 10.2% of the average team's at bats by Pujols, we find the marginal impact is that Pujols would increase the number of runs scored for an average team from 783 to 853. How many wins is that worth? With Pujols our scoring ratio is $853/783 = 1.089$. Using the Pythagorean Theorem from chapter 2 we predict that the team with Pujols would win $\dfrac{162 \times (1.089)^2}{(1 + 1.089^2)} = 87.38$ games. Therefore, we would estimate that adding Pujols to an average team would lead to $87.38 - 81 = 6.38$ wins. We will see in chapter 9 that an alternative analysis

	H	I	J
2	**Outcome**	**Number**	**Probability**
3	Plate Appearances	6236.27	
4	At Bats +Sac. Hits + Sac. Bunts	5658.03	
5	Errors	102	0.01635593
6	Outs (in play)	3027.23	0.48542318
7	Strikeouts	1026.37	0.16458075
8	BB	528.23	0.08470288
9	HBP	50	0.00801761
10	Singles	986.67	0.15821477
11	2B	304.5	0.04882726
12	3B	31.73	0.00508798
13	HR	179.53	0.02878804

Figure 4.8. Average team statistics for the MLB, 2006. See file Pujolsoveraverage.xls.

of Pujols's 2006 batting record indicates that he added around 9.5 wins more than an average player would have.

APPENDIX

Using a Data Table to Perform a Simulation in Excel

In the cell range B2:H22 of the file simulationmotivator.xls we have programmed Excel to "play out" an inning for a team in which each hitter has a 50% chance of striking out or hitting a home run. Hit F9 and the number of runs scored by the team is recorded in cell H3. Note that whenever the Excel RAND() function returns a value less than 0.5 the batter hits a home run; otherwise the batter strikes out. To record the number of runs scored during many (say, 1,000) innings, we enter the numbers 1 through 1,000 in the cell range I6:I1005. Next we enter in cell J3 the formula (=H3) that we want to play out or simulate 1,000 times. Now we select the cell range I5:J1005 (this is called the Table Range). In Excel 2003 or earlier select Data Table. In Excel 2007 select Data and then choose the What-If icon (the one with a question mark) and choose Data Table.

Next leave the row input cell blank and choose any blank cell as your column input cell. Then Excel puts the numbers 1, 2, . . . 1,000 successively in your selected blank cell. Each time cell H3 (runs in the inning) is

	I	J
2		**Average Runs per Inning**
3		3.00
4		
5		4
6	1	0
7	2	2
8	3	0
9	4	3
10	5	0
11	6	4
12	7	4
13	8	6
14	9	0
15	10	0
16	11	3
1003	998	2
1004	999	2
1005	1000	7

Figure 4.9. Simulating 1,000 innings of Joe Hardy hitting.

recalculated as the RAND() functions in column C recalculate. Entering the formula =AVERAGE(I6:I1005) in cell J3 calculates the average number of runs scored per inning during our 1,000 simulated innings. For the 1,000 innings simulated in figure 4.9, the mean number of runs scored per inning was 3.

EVALUATING BASEBALL PITCHERS AND FORECASTING FUTURE PERFORMANCE

In chapters 2–4 we discussed three methods that can be used to evaluate the performance of a baseball hitter: Runs Created, Linear Weights, and Monte Carlo simulation. Let's turn our attention to evaluating the performance of baseball pitchers. As we will see, evaluating their performance is no easy matter.

Until recently, the most frequently used technique for evaluating the performance of pitchers was earned run average (ERA). Let's consider a pitcher, again named Joe Hardy. Consider all the runners Joe allows to reach base. Any of the base runners who score or would have scored if Joe's team made no fielding miscues (such as an error, passed ball, and so forth) causes Joe to be charged with an earned run. For example, if Joe gives up a triple with two outs in an inning and the next batter hits a single and a run is scored, Joe is charged with an earned run. Now suppose instead of a single the next batter hits a ball to the shortstop, who misplays the ball and is charged with an error. If the runner scores, this is an unearned run because without the error the runner would not have scored. A pitcher's ERA is the number of earned runs he gives up per nine innings. For example, if Joe gives up 20 earned runs in 45 innings, he has given up $\frac{20 \times 9}{45} = 4$ runs per nine innings and thus his ERA is 4. In general, a pitcher's ERA is computed as $\frac{(\text{earned runs allowed}) \times 9}{\text{innings pitched}}$.

Problems with ERA

There are several problems with evaluating pitchers by their ERA.

1. Errors are subjective. Some official scorers are more reluctant than others to call a batted ball an error. David Kalist and David Spurr have

found slight evidence that official scorers are biased in favor of the home team.[1]

2. When a starting pitcher is pulled from the game and there is at least one base runner, the number of earned runs he gives up depends greatly on the performance of the relief pitcher. For example, suppose that Joe leaves the game with two outs and the bases loaded. If the relief pitcher gets the next out, Joe is charged with no earned runs, but if the relief pitcher gives up a grand slam, then Joe is charged with three earned runs.

3. A pitcher with good fielders behind him will clearly give up fewer earned runs than a pitcher with a leaky defense. (We will discuss the evaluation of fielders in chapter 7.)

Starting pitchers are often evaluated on the basis of their win-loss record. This clearly depends on the batting support the pitcher receives. For example, in 2006 the great Roger Clemens had an ERA of 2.30 (approximately half the league average), but he had a 7–6 record because he received poor batting support from the Houston Astros.

Relief pitchers are often evaluated on the basis of how many saves they have in a given season. Most saves credited to relief pitchers are given to a relief pitcher who faces a batter representing the tying run. The following extract provides the official definition of a save.

> The official scorer shall credit a pitcher with a save when such pitcher meets all four of the following conditions:
>
> (1) He is the finishing pitcher in a game won by his team;
> (2) He is not the winning pitcher;
> (3) He is credited with at least a third of an inning pitched; and
> (4) He satisfies one of the following conditions:
>   (a) He enters the game with a lead of no more than three runs and pitches at least one inning;
>   (b) He enters the game, regardless of the count, with the potential tying run either on base, or AB or on deck; or
>   (c) He pitches for at least three innings.[2]

To paraphrase George Orwell, "All saves are created equal, but some saves are more equal than others." In looking at a relief pitcher's published

---

[1] Kalist and Spurr, "Baseball Errors."

[2] The official definition of a save has been reprinted by special permission of the Office of the Commissioner of Baseball from the *Official Baseball Rules*. The copyright in the *Official Baseball Rules* is owned and has been registered by the Commissioner of Baseball.

statistics, all we see is the number of saves, so all saves seem equal. However, consider a relief pitcher who comes in with his team ahead 3–2 during the top of the ninth inning and other team has bases loaded and none out. If this pitcher holds the lead, he has done a fabulous job. Consider a second relief pitcher who enters the game with two outs in the ninth and a runner on first and a 4–2 lead. He strikes out the next batter and receives a save. Clearly the first relief pitcher deserves much more credit, but each pitcher receives a save.

In chapter 8 we will show how Player Win Averages resolve many of the problems involved in evaluating pitcher performance. Player Win Averages also allow us to compare the value of relief pitchers and starting pitchers.

Using Past ERA to Predict Future ERA
Does Not Work Well

Despite the problems with ERA, it seems important to be able to predict a pitcher's future ERA from his past performance. This would aid baseball management in their quest to improve their team's future pitching performance.

It seems logical to try to predict a pitcher's ERA for the next season from his previous season's ERA. For a long time the baseball community thought that this approach would yield good predictions of the following year's ERA. Let's check out this hypothesis. For all pitchers who pitched at least 100 innings during two consecutive seasons in 2002–6, figure 5.1 plots on the x axis the pitcher's ERA during a given season and on the y axis the pitcher's ERA during the following season. I used Excel's Trend Curve Feature to plot the line (see the chapter appendix for details on how to use the Trend Curve feature) that best fits these data.

Figure 5.1 indicates that the best-fitting[3] straight line that can be used to predict the following year's ERA by using the previous year's ERA is the equation

$$(\text{next year's ERA}) = 2.8484 + .353(\text{last year's ERA}). \qquad (1)$$

For example, a pitcher who had an ERA of 4.0 in a given year would be predicted to have an ERA the following year of $2.8484 + .353(4) = 4.26$.

We can see from figure 5.1 that the best-fitting line does not fit the data very well. Many pitchers with predicted ERAs of around 4.0 actually have ERAs the following year of over 6.0 or less than 2.0. Statisticians quantify

---

[3] Excel chooses the line as best fitting which minimizes the sum of the squared vertical distances of the points to the fitted line. This is called the least squares line.

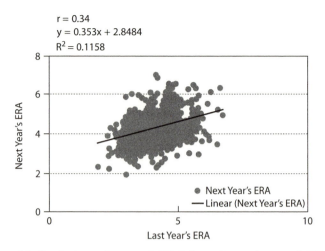

Figure 5.1. Predicting a given year's ERA from the previous year's ERA.

how well a line fits a set of data using the R Squared value (RSQ) and correlation as well as the mean absolute deviation of the regression forecasts.

R² and Correlation

From figure 5.1 we find that the RSQ for predicting the next year's ERA from the previous year's ERA is .116. This indicates that the previous year's ERA explains only 11.6% of the variation in the following year's ERA. In other words, 88% of the variation in the following year's ERA is unexplained by the previous year's ERA. Statisticians also measure linear association by looking at the square root of RSQ, which is often called r or the correlation coefficient. We find that the correlation between the two years' ERA is $\sqrt{.116} = .34$.

The correlation (usually denoted by r) between two variables (X and Y) is a unit-free measure of the strength of the linear relationship between X and Y. The correlation between any two variables is always between -1 and $+1$. The exact formula used to compute the correlation between two variables is not very important.[4] It is important, however, to be able to interpret the correlation between X and Y.

A correlation near $+1$ means that there is a strong positive linear relationship between X and Y. That is, when X is larger than average Y tends

---

[4] If you wish, however, you can use Excel to compute correlations between two columns of numbers with the =CORREL function. See the chapter appendix for details.

to be larger than average, and when X is smaller than average Y tends to be smaller than average. Alternatively, when a straight line is fit to the data, there will be a straight line with positive slope that does a good job of fitting the points. As an example, for the data shown in figure 5.2 (here X = units produced and Y = cost), X and Y have a correlation of +0.90.

A correlation near −1 means that there is a strong negative linear relationship between X and Y. That is, when X is larger than average Y tends to be smaller than average, and when X is smaller than average Y tends to be larger than average. Alternatively, when a straight line is fit to the data, there will be a straight line with negative slope that does a good job of fitting the points. For the data shown in figure 5.3 (X = price and Y = demand), X and Y have a correlation of −0.94.

A correlation near 0 means that there is a weak linear relationship between X and Y. That is, knowing whether X is larger or smaller than its mean tells

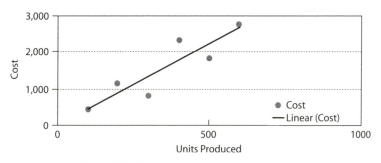

Figure 5.2. Strong positive linear relationship.

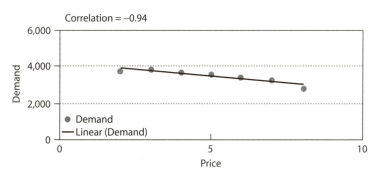

Figure 5.3. Strong negative linear relationship.

Figure 5.4. Weak linear relationship.

you little about whether Y will be larger or smaller than its mean. Figure 5.4 shows the dependence of Y (sales) on X (years of experience). Years of experience and unit sales have a correlation of .003. In our data set average experience is 10 years. We see that when a person has more than 10 years of sales experience, his sales can either be low or high. We also see that when a person has less than 10 years of sales experience, sales can be low or high. Although there is little or no linear relationship between experience and sales, we see there is a strong non-linear relationship (see fitted curve) between years of experience and sales. Correlation does not measure the strength of non-linear relationships.

Mean Absolute Deviation of Forecasts

The average absolute deviation (MAD for short) of the forecast errors is a commonly used measure of forecast accuracy. For each pitcher we compute the predicted ERA from (1) and take the absolute value of (predicted ERA) − (actual ERA). We find the MAD for predicting the following year's ERA from the previous year's ERA to be 0.68 runs. In other words, our error on average in predicting the following year's ERA from the previous year's ERA is 0.68 runs.

Voros McCracken Stuns the Baseball World

Voros McCracken appears to be the first person to successfully explain why future ERAs are hard to predict from past ERAs.[5] McCracken observed that a pitcher's effectiveness is primarily based on the following:

[5] See Voros McCracken, "Pitching and Defense: How Much Control Do Hurlers Have?" January 23, 2001, http://www.baseballprospectus.com/article.php?articleid=878.

1. The fraction of BFP (Batters Faced by Pitchers) that result in balls in play (a ball in play is a plate appearance that results in a ground out, error, single, double, triple, fly out, or line out).

2. The fraction of balls in play that result in hits (referred to as BABIP, or Batting Average on Balls in Play).

3. The outcome of BFP that do not result in balls in play. What fraction of BFP that do not yield a ball in play result in strikeouts, walks, HBP, or home runs?

McCracken's brilliant insight was that a pitcher's future performance with regard to the situations outlined in (1) and (3) can be predicted fairly well from past performance, *but it is very difficult to predict (2) from past performance.* Suppose we try to predict the percentage of BFP resulting in strikeouts for a pitcher using the percentage of BFP resulting in strikeouts during his previous season. We find r = .78. For BB, r = .66 and for HR, r = .34. McCracken called SO, BB, HBP, and HR Defense Independent Pitching Statistics (DIPS for short) because these results are independent of the team's fielding ability. DIPS seem to be fairly predictable from season to season. However, the fraction of balls in play resulting in an out or a hit seems to be very hard to predict.[6] For example, a pitcher's BABIP has only a .24 correlation with a pitcher's BABIP for his previous season. The unpredictability of BABIP is what makes it so difficult to predict a pitcher's ERA in a given season using his ERA from the previous year. McCracken sums things up in the following way: "The pitchers who are the best at preventing hits on balls in play one year are often the worst at it the next. In 1998, Greg Maddux had one of the best rates in baseball, then in 1999 he had one of the worst. In 2000, he had one of the better ones again. In 1999, Pedro Martinez had one of the worst; in 2000, he had the best. This happens a lot."[7]

I believe that luck and season-to-season differences in team fielding quality are major factors in the lack of predictability of BABIP. Much research needs to be done in this area, however.

DICE: A Better Model for Predicting a Pitcher's Future Performance

So how can we use McCracken's insights to better predict a pitcher's future ERA? McCracken came up with a very complex method to predict future

---

[6] Later researchers found that for certain types of pitchers (particularly knuckle ball pitchers) the outcome of balls in play is much easier to predict.

[7] McCracken, "Pitching and Defense."

ERA. Since DIPS are fairly predictable from year to year, it seems reasonable that there should be some simple combination of DIPS (BB, SO, HBP, and HR) that can be used to predict ERA more accurately than our previous approach. In 2000 Clay Dreslogh came up with a simpler formula, known as Defense-Independent Component ERA (DICE), to predict ERA.[8]

$$DICE = 3.00 + \frac{13HR + 3(BB + HBP) - 2K}{IP} \tag{2}$$

As we can see from equation (2), DICE predicts ERA by plugging in a pitcher's HR, K, BB, HBP, and Innings Pitched (IP) from the previous year. For example, in 1997 Roger Clemens had the following statistics:

- 68 BB
- 7 HBP
- 292 K
- 9 HR
- 264 IP

Using these statistics in equation (2), we predict Clemens's ERA for 1998 to be

$$DICE = 3.00 + (3(68 + 7) + (13 \times 9) - (2 \times 292) / 264 = 2.08.$$

Clemens's actual ERA in 1998 was 2.05.

For the years 2001–6 for all pitchers who completed 100 or more innings in consecutive seasons, I computed each pitcher's DICE for year x and used this to try and predict the ERA for year x + 1. For example, for a pitcher who pitched over 100 innings in 2003 and 2004, we would use the pitcher's 2003 DICE to predict the pitcher's 2004 ERA. The results are shown in figure 5.5.

Therefore, we predict a given year's ERA as .56(last season's DICE) + 1.975. The previous year's DICE explains 19% of the variation in the following year's ERA (compared to the previous year's ERA, which explained only 11% of the following year's ERA). The correlation between the previous year's DICE and the following year's ERA is .44 (the previous year's ERA had only a .34 correlation with the following year's

[8] Clay Dreslogh, "DICE: A New Pitching Stat," July 19, 2000, http://www.sportsmogul.com/content/dice.htm.

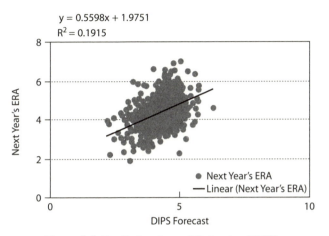

$y = 0.5598x + 1.9751$
$R^2 = 0.1915$

Figure 5.5. Predicting future ERA using DICE.

ERA). On average, our predictions of a given year's ERA from the previous year's DICE were off by only 0.51 runs (predictions based on last year's ERA were off by 0.68 runs).

To sum up, we can more accurately predict a given year's ERA from HR, BB, HBP, and K than from a previous year's ERA. In chapter 8 we will use Player Win Averages to come up with a better way to evaluate a pitcher's performance.

A Holy Grail of Mathletics

In this chapter we have briefly discussed the problems involved in developing a model to predict a baseball pitcher's future performance. There are many other important sports problems that involve predicting a player's or team's future performance from past performance. These include predicting

- the performance of an NBA, NFL, or MLB draft pick from his high school, college, or international performance;
- a batter's future Runs Created;
- a running back's, quarterback's, or wide receiver's future performance from his past professional performance;
- a team's record for a given season based on past-season performance and player trades.

Many people have developed such forecasting models. For example, each season Ron Shandler's *Baseball Forecaster*, Bill James's annual *Baseball Handbook*, and Baseball Prospectus's *Baseball Annual* issue projections for each MLB team regarding their performance in the following season. Each year Aaron Schatz's *Football Prospectus* (and dozens of fantasy football magazines) predict NFL team and player performance for the following season. Each year ESPN.com's John Hollinger predicts NBA team and player performance for the following season.

Though these forecasts are fascinating, what is really needed is a comparison of the forecasting accuracy of various methods. Then we can judge which forecasts to use in our fantasy leagues, draft decisions, or player personnel decisions. For example, who predicts future baseball performance better: Ron Shandler or Bill James? Perhaps in the future databases of accuracy of sports forecasts (and stock pickers) will be commonplace on the Internet.

APPENDIX

Using the Excel Trend Curve

The Excel Trend Curve feature enables us to plot the line that best fits a set of data. In the file trendcurveexample.xls we are given the ERA of pitchers who pitched over 100 innings during both the 2005 and 2006 seasons. We would like to plot on the x axis each pitcher's 2005 ERA and on the y axis the same pitcher's 2006 ERA. Then we want to graph the line (and obtain the equation of the line) that best fits this relationship.

Excel 2003 or Earlier

Select the data to be graphed (cell range D4:E98). Then select the Chart Wizard icon (it looks like a chart) and choose XY scatter and select the first option. Then click on the graphed points until they turn yellow. After right clicking on any of the points, select Add Trendline and choose the Linear Option. Then check the Display Equation and Display R Sq options. You will then see the graph shown in figure 5.6.

We find that the straight-line relationship that best predicts a pitcher's ERA during 2006 is 2006 ERA = .3802(2005 ERA) + 2.75. This equation explains 11.1% of the variation in 2006 ERA.

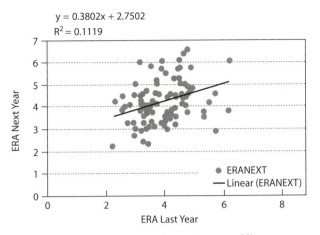

Figure 5.6. Example of creating a trend line.

Excel 2007

Select the data to be graphed (cell range D4:E98). Then select Insert and choose the first option from Scatter. Then click on the graphed points until they turn blue. After right clicking on any of the points, select Add Trendline and choose the Linear Option. Then check the Display Equation and Display R Sq options. You will then see the graph shown in figure 5.6.

We find that the straight-line relationship that best predicts a pitcher's ERA during 2006 is 2006 ERA = .3802(2005 ERA) + 2.75. This equation explains 11.1% of the variation in 2006 ERA.

6

BASEBALL DECISION-MAKING

During the course of a season, managers make many crucial decisions, including the ones listed below.

- With a man on first and nobody out should we attempt a sacrifice bunt to advance the runner to second base?
- With a man on first and one out should we attempt to steal second base?
- We are the home team and the score is tied in the top of the ninth inning. The opposing team has a man on third base and none out. Should we play the infield in? This increases the chance of a hit (most people think bringing the infield in makes a .250 hitter a .300 hitter), but bringing the infield in ensures that a ground out will not score the runner from third.

Decision-making in baseball, as in all aspects of life and business, involves making trade-offs. Let's analyze the decision concerning whether to try to bunt with a runner on first and none out. If the bunt succeeds, the runner will advance to second and be one base closer to scoring, but a precious out will have been given up. Is the benefit of the extra base worth giving up the out? We will soon see that in most situations, the benefit obtained from advancing the runner to second base does not justify relinquishing an out.

Possible States during a Baseball Game

The key to developing a framework for baseball decision-making is to realize that during an inning a team is in one of the twenty-four situations (often called states) listed in table 6.1.

Each state is denoted by four numbers. The first number is the number of outs (0, 1, or 2). The second number lets us know if first base is occupied (1 = base occupied, 0 = base not occupied). Similarly, the third and fourth numbers tell us whether second or third base, respectively, are occupied. For example, 1010 means there is one out and a runner on second

TABLE 6.1
Possible States during an Inning

State	Outs	Runner on First?	Runner on Second?	Runner on Third?
0000	0	No	No	No
1000	1	No	No	No
2000	2	No	No	No
0001	0	No	No	Yes
1001	1	No	No	Yes
2001	2	No	No	Yes
0010	0	No	Yes	No
1010	1	No	Yes	No
2010	2	No	Yes	No
0011	0	No	Yes	Yes
1011	1	No	Yes	Yes
2011	2	No	Yes	Yes
0100	0	Yes	No	No
1100	1	Yes	No	No
2100	2	Yes	No	No
0101	0	Yes	No	Yes
1101	1	Yes	No	Yes
2101	2	Yes	No	Yes
0110	0	Yes	Yes	No
1110	1	Yes	Yes	No
2110	2	Yes	Yes	No
0111	0	Yes	Yes	Yes
1111	1	Yes	Yes	Yes
2111	2	Yes	Yes	Yes

base, and 2001 means there are two outs and a runner on third base. Intuitively we know the best state is 0111 (bases loaded nobody out) and the worst state is 2000 (two outs nobody on). How can we explicitly measure how much better one state is than another? We simply look at the average number of runs scored in each situation over the course of many games. In *Baseball Hacks*, Joseph Adler has tabulated (see table 6.2) the average number of runs scored in each situation based on data from the 2004 season.

For example, with a runner on first and second and one out (state 1110), teams scored an average of .97 runs. This situation occurred 4,978 times. The information in table 6.2 is vital for proper baseball decision-making. To see why, let's look at state 0100 (a runner on first and none out). Since our table aggregates data over all teams and batters, we will assume the numbers in table 6.2 refer to the expected number of runs scored given that "an average" hitter is at bat. In state 0100 we see that an "average" team is expected to score .93 runs. Now if a sacrifice bunt is attempted, it might succeed in advancing the runner on first to second, resulting in the batter being out (leading us to state 1010), or fail by having the lead runner out and the batter reaching first (state 1100). These are by far the most common outcomes when a bunt is attempted. If, on average, more runs are scored with a bunt than without a bunt, then bunting is a good idea. How do we figure out the average number of runs a team will score after bunting? Before proceeding further, a brief introduction to some important concepts in mathematical probability theory is in order.

Experiments and Random Variables

First let's define the important concepts of experiment and random variable. An experiment is any situation whose outcome is uncertain. Examples of experiments include

- a dice throw (outcomes of 1, 2, 3, 4, 5, or 6 are possible for each die)
- a batter's plate appearance (the many possible outcomes include a home run, a single, a strikeout, and so forth)
- a free throw (outcomes include successful or unsuccessful free throw; rebound by either team)
- a pass thrown by a quarterback (outcomes include incomplete pass, interception, completion for 10 yards, completion for 15 yards, and so forth)
- a field goal attempt (outcome is either a made or missed field goal)

Random variables can be associated with experiments. Here are some examples:

TABLE 6.2
Expected Runs

State	Average Runs	Number of Plate Appearances for This Situation
0000	.54	46,180
1000	.29	32,821
2000	.11	26,009
0001	1.46	512
1001	.98	2,069
2001	.38	3,129
0010	1.17	3,590
1010	.71	6,168
2010	.34	7,709
0011	2.14	688
1011	1.47	1,770
2011	.63	1,902
0100	.93	11,644
1100	.55	13,483
2100	.25	13,588
0101	1.86	1,053
1101	1.24	2,283
2101	.54	3,117
0110	1.49	2,786
1110	.97	4,978
2110	.46	6,545
0111	2.27	805
1111	1.6	1,926
2111	.82	2,380

Source: Joseph Adler, *Baseball Hacks* (O'Reilly Media, 2006), 313.

- the sum of the total on two dice (possible values include 2, 3, . . . 10, 11, 12)
- the number of runners batted in during a batter's plate appearance (0, 1, 2, 3, or 4)
- the number of points scored on free throw (possible values are 0 and 1)
- the number of points scored on passing play (0 and 6 are possible)
- the number of points scored on a field goal (0 and 3 are possible)

Expected Value

In our analysis of baseball, basketball, and football, we will often need to determine the expected value of a random variable. The expected value of a random variable is the average value of the random variable we can expect if an experiment is performed many times. In general we find the expected value of a random variable as follows:

$$\sum_{\text{all outcomes}} (\text{probability of outcome}) \times (\text{value of random variable for outcome}).$$

For example, if we toss a die, each possible outcome has probability of 1/6. Therefore, if we define the random variable X = Number of dots showing up when die is tossed, then

$$E(X) = \text{expected value of } X = \frac{1}{6}(1) + \frac{1}{6}(2) + \frac{1}{6}(3)$$
$$+ \frac{1}{6}(4) + \frac{1}{6}(5) + \frac{1}{6}(6) = 3.5.$$

Therefore, if we were to toss a die many times and average the total number of dots, we would expect to get a number near 3.5.

In baseball, we will compare various decisions based on expected runs scored. For example, if expected runs scored is higher if a team bunts than if it doesn't, then the team should not bunt. In football or basketball we will usually compare decisions based on expected number of points by which a team beats an opponent. For example, suppose in football a team faces fourth and 3 on its opponent's 35-yard line. If a field goal means the team beats the opposition on average by 0.5 points during the rest of the game and going for the first down means the team wins by an average of

1.5 points during the rest of the game, then the team should eschew the field goal and go for the first down.[1]

In computing expected values of random variables we will often use the Law of Conditional Expectation.

expected value of random variable

$$= \sum_{\text{all outcomes}} (\text{probability of outcome}) \times (\text{expected value of}$$

random variable given outcome).

For example, suppose in football a running play gains an average of five yards if an opponent plays a passing defense and an average of three yards if an opponent plays a rushing defense. Also assume the opponent plays a rushing defense 40% of time and passing defense 60% of time. Then we can use the Law of Conditional Expectation to compute the expected number of yards gained on a running play as

(probability of pass defense)
× (expected yards gained given pass defense is played)
+ (probability of run defense)
× (expected yards gained given run defense is played)
= (.6)(5) + (.4) × (3) = 4.2 yards per play.

To Bunt or Not to Bunt—That Is the Question

We are now ready to determine whether bunting with a man on first and nobody out is a good play. Adler tabulated (for the 2004 season) the results of bunts with a runner on first and found the results shown in table 6.3.

From the data in table 6.2 we know that in the current state (0100), the team will score on average .93 runs. Since this is based on data from all teams and players, this number essentially assumes an average batter is at the plate. If a great hitter is up, the expected runs would be more than .93 runs while if a poor hitter is up the expected runs would be fewer than .93 runs.

---

[1] If it is near the end of the game, however, the team should maximize the probability of winning the game. If it is not near end of game, then maximizing choosing decisions based on maximizing expected number of points by which a team beats an opponent is virtually equivalent to maximizing the probability of victory. We will use this approach in chapter 22 to study the basis for making important football decisions such as whether a team should go for a field goal on fourth down.

TABLE 6.3
Possible Results of a Bunt with Runner on First

Result	Resulting State	Probability	Expected Runs*
Batter is safe and runner advances to second base	0110	.10	1.49
Runner advances to second base and batter is safe	1010	.70	.71
Both runners are out	2000	.02	.11
Runner is out at second base and batter reaches first base	1100	.08	.55
Batter is out and runner remains on first base	1100	.10	.55

*Expected runs are derived from data in table 6.2.

Applying the Law of Conditional Expectations to the data in table 6.3, we find the expected number of runs scored after the bunt is

$$.10(1.49) + .70(.71) + .02(.11) + .08(.55) + .10(.55) = .75 \text{ runs.}$$

Thus bunting makes a team, on average, .18 runs ($.75 - .93 = -.18$) worse-off than if it does not bunt. Thus, bunting is not a good idea if an average hitter is up and the goal is to maximize the expected number of runs in an inning.

What If the Batter Is a Poor Hitter?

What if a really bad hitter is up? Let's assume a weak-hitting pitcher, Joe Hardy, is up. Joe strikes out 85% of the time, hits a single 10% of the time, and walks 5% of the time. We will assume the single always advances a runner on first to third base. If we do not bunt with Joe at the plate, the Law of Conditional Expectation tells us the expected number of runs we will score in the inning is given by

$$.85 \times E(1100) + .10 \times E(0101) + .05E(0110),$$

where E(state) is expected runs scored in that state. Using the expected runs for each state from table 6.2 we find that with our weak hitter up we can expect to score

$$.85(.55) + .10(1.86) + .05(1.49) = .73 \text{ runs.}$$

Therefore, for Joe Hardy, bunting would actually increase the number of expected runs.

Is Bunting a Good Idea with the Score Tied?

Is bunting a good idea when a team needs to score only one run? For example, suppose the score is tied in the bottom of the ninth inning and a team has a runner on first with none out. If the team scores a run they win the game. Should they bunt? We know that unless a very weak hitter is up, bunting will decrease the expected number of runs scored. In this situation, however, we want to look at *the probability of scoring at least one run*. In the excellent book *Baseball between the Numbers*, the authors tabulate the probability of scoring at least one run for all twenty-four states. With a runner on first and none out, the probability of scoring at least one run is .417. The other probabilities that are germane to our example are summarized in table 6.4.

The Law of Conditional Expectation tells us that bunting will yield a probability

$$.10(.625) + .70(.41) + .02(.071) + .08(.272) + .10(.272) = .40$$

of scoring at least one run. Therefore bunting and not bunting give us just about the same probability of scoring at least one run. Therefore if an average hitter is at bat, a team should be indifferent regarding bunting and not bunting.

To Steal or Not to Steal (a Base)

Let's now examine the stolen base decision. Let's suppose we have a runner on first base and none out. Let p = probability of a successful steal of second base. For which values of p should we steal? In our current state (0100) we expect to score .93 runs. If the steal is successful, the new state is 0010, in which we expect to score 1.17 runs. If steal is unsuccessful, the new state is 1000, in which we expect to score 0.29 runs. Thus if we steal, the Law of Conditional Expectations tells us our expected runs scored is $1.17p + .29(1-p)$. As long as $1.17p + .29(1-p) > .93$ we should steal

TABLE 6.4
Probabilities of Scoring at Least One Run

Result	Resulting State	Probability	Probability of Scoring at Least One Run
Batter is safe and runner advances to second base	0110	.10	.625
Runner advances to second base and batter is safe	1010	.70	.41
Both runners are out	2000	.02	.071
Runner is out at second base and batter reaches first base	1100	.08	.272
Batter is out and runner remains on first base	1100	.10	.272

Source: Baseball Prospectus Team of Experts, Jonah Keri, and James Click, *Baseball between the Numbers: Why Everything You Know about the Game Is Wrong* (Perseus Publishing, 2006), 129.

second base. Solving this inequality, we find we should steal if $.88p > .64$ or $p > .64/.88 = .727$. Therefore, if our chance of stealing second base exceeds 72.7%, then trying to steal second base is a good idea. Over the last seven years, 70% of all stolen base attempts have been successful, which indicates that teams try to steal more often than they should.

In a similar fashion we find that trying to steal second base with one out increases expected runs if the probability of success is at least 75%. In a similar fashion we find that trying to steal second base with two outs increases expected runs if the probability of success is at least 73.5%. Stealing third base with none out increases expected runs if the probability of success is at least 75.2%, while with one out a team needs only the chance of a successful steal for the probability of success to exceed 69.2%. With two outs the probability of a successful steal of third base needs to be at least 89.4% to make it worthwhile.

Are Base Runners Too Conservative When Trying to Advance on a Single or Double?

If there is a runner on first and a single is hit, the coaches and runner must decide whether to try and advance to third base or stop at second base. If a runner is on second and a single is hit, the runner and coaches must decide whether to stop at third base or try to score. If a runner is on first and a double is hit, the runner and coaches must decide whether to try to score or stop at third base. As you'll see below, most major league teams are much too conservative when deciding whether the base runner should try and "go for the extra base."

Let's suppose a team has a runner on first and nobody out. If the next batter hits a single and the runner makes it to third base, we know from table 6.2 that the team is in a situation (first and third, nobody out) worth on average 1.86 runs. If the runner is out, there will be a runner on first with one out (assuming the batter does not take second on the throw). In this situation the team averages scoring 0.55 runs. If the runner stops at second, there will be runners on first and second with none out, which yields an average of 1.49 runs. Let p = the probability the runner makes it from first to third. The team will maximize its expected runs by trying to advance if and only if $p(1.86) + (1 - p).55 \geq 1.49$. We find this inequality is satisfied for $p \geq .72$. This implies that a runner on first with nobody out should try to go for third if his chance of success is at least 72%. According to data from the 2005 MLB season,[2] base runners trying to go from first to third are thrown out only 3% of the time. This means there are probably many situations in which base runners had an 80–90% chance of making it from first to third in which they did not try to advance (and they should have). The "breakeven" probability of successfully taking the extra base that is needed to justify the attempt to do so is summarized in table 6.5.

The probability of being thrown out when trying to score on a single from second base is around 5%, so with none out runners are behaving in a near optimal fashion because they should advance as long as their chance of being thrown out is less than 5%. In all other situations, however, runners are not trying to advance in many situations in which they should be. For example, a runner on second with two outs should try to score as long he has at least a 43% chance of scoring.

---

[2] See Jeff Angus, "Can Baserunning Be the New Moneyball Approach?" http://base ballanalysts.com/archives/2005/10/can_baserunning.php.

TABLE 6.5
Breakeven Probability Needed to Justify Trying for
the Extra Base

Runner on	Number of Outs	Breakeven Probability of Success Needed on a Single
First	0	.72
First	1	.73
First	2	.85
Second	0	.95
Second	1	.76
Second	2	.43

Note: See file baserunners.xls for the calculations.

In summary, when deciding between strategic options such as bunting or not bunting we choose the strategy that yields the largest number of expected runs or (if the game is tied late) maximizes the probability of scoring at least one run. As a general rule we could choose the decision that maximizes the expected probability of winning the game. In most situations, a decision that maximizes expected runs scored will also maximize the chance of winning the game. The determination of the probability of winning a game in any game situation is discussed in chapter 8 and after reading this chapter the reader can, if she desires, make decisions based on maximizing the probability of winning the game. This approach will always yield the correct decision, but it is much more difficult to implement than maximizing expected runs. Therefore, in this chapter we have chosen to describe how to maximize expected runs.

APPENDIX

Runs Per Inning

Recall in chapter 4 we stated (without proof) and verified by simulation that a team in which each batter had a 50% chance of hitting a home run and a 50% chance of striking out would average 3 runs per inning. We now use conditional expectation to prove this result.

Let R_i = expected runs scored by this team in an inning in which i outs are allowed. Then $R_1 = .5(0) + .5(1 + R_1)$. This follows because with probability .5 the first batter makes an out and the inning ends with the team scoring 0 runs. Also with probability .5 the first batter hits a home run and the team can expect to score $1 + R_1$ runs, because there is still one out left. Solving this equation we find $R_1 = 1$.

Now we can solve for R_2 from the following equation:

$$R_2 = .5(R_1) + .5(1 + R_2).$$

This equation follows because with probability .5 the hitter makes an out and the team can expect to score R_1 runs. Also with probability .5 he hits a home run and the team can expect to score a total of $1 + R_2$ runs because there are 2 outs remaining. After substituting in that $R_1 = 1$ we find that $R_2 = 2$.

Now we can solve for R_3 using the equation

$$R_3 = .5(R_2) + .5(1 + R_3).$$

After substituting in that $R_2 = 2$, we find that $R_3 = 3$, as we claimed. Generalizing this logic we can easily show that $R_n = n$.

EVALUATING FIELDERS

Sabermetrics' Last Frontier

Surprisingly, until the late 1990s little progress was made in determining how to evaluate the effectiveness of fielders and the relative importance of fielding (as compared to batting and hitting). Until recently the prevailing wisdom in baseball was that you had to have "strength up the middle" (good fielding at second base, shortstop, catcher, and center field) to have a good team. We will see that in most cases, the differences in player fielding abilities are not significant enough to be a major factor in team performance. As the saying goes, the exception proves the rule and we will see that the 2005 Yankees were a very poor fielding team, and we can estimate that their poor fielding cost them approximately eleven wins.

Fielding Percentage: The Traditional, Fatally Flawed Metric

In our discussion of fielding, we will focus primarily on the important position of shortstop (SS). Until recently, the only measure of fielding effectiveness available was Fielding Percentage. For any fielder,

$$\text{fielding percentage} = \frac{PO + A}{PO + A + E}.$$

PO = putouts made by the fielder. For example, a SS gets credit for a putout when he catches a fly ball or line drive, tags a runner out, or receives the ball and steps on second base to complete a force out.

A = assists made by the fielder. For example, a SS gets credit for an assist when he throws to first base and the batter is put out.

E = errors made by the fielder. Again, whether a batted ball is scored an error is a subjective decision made by the official scorer.

	C	D	E	F	G	H	I
13	Year	InnOuts	Putouts	Assists	Errors	Range Factor	Fielding %
14	2000	3836	236	349	24	0.908110132	0.960591133
15	2001	3937	211	344	15	0.83943838	0.973684211
16	2002	4150	219	367	14	0.8480834984	0.976666667
17	2003	3101	160	271	14	0.827631015	0.968539326
18	2004	4025	273	392	13	0.983823097	0.980825959
19	2005	4058	262	454	15	1.050660084	0.979480164
20	2006	3877	214	381	15	0.913865786	0.975409836

Figure 7.1. Derek Jeter's fielding statistics, 2000–2006. InnOuts = the number of defensive outs for which Jeter was on the field; PO = putouts made by the fielder; A = assists made by the fielder; E = errors made by the fielder; RF = Range Factor; FP = Fielding Percentage.

	D	E	F	G	H	I	J
16	Year	InnOuts	Putouts	Assists	Errors	Range Factor	Fielding %
17	2000	2518	147	289	23	1.031079179	0.949891068
18	2001	2083	126	224	11	1.000552837	0.969529086
19	2002	3943	245	466	27	1.073752226	0.963414634
20	2003	4050	237	481	31	1.055676064	0.958611482
21	2004	3402	192	411	24	1.055466021	0.961722488
22	2005	3919	255	504	15	1.153261426	0.980620155
23	2006	4113	269	492	27	1.101760507	0.965736041

Figure 7.2. Rafael Furcal's fielding statistics, 2000–2006. InnOuts = the number of defensive outs for which Furcal was on the field; PO = putouts made by the fielder; A = assists made by the fielder; E = errors made by the fielder; RF = Range Factor; FP = Fielding Percentage.

Essentially, Fielding Percentage computes the percentage of balls in play that a fielder handles without making an error. Figures 7.1 and 7.2 give fielding data for two shortstops: Derek Jeter and Rafael Furcal during the 2000–2006 seasons. Most casual baseball fans think Jeter is a great fielder. As we will soon see, this is not the case.

To illustrate the computation of Fielding Percentage (listed in the FP column), let's compute Jeter's 2004 Fielding Percentage: $FP = \dfrac{273+392}{273+392+13} = .981$. Therefore, Jeter properly handled 98.1% of his fielding chances. During the years 2000–2006 the average Fielding Percentage for a SS was .974, so Jeter's 2004 performance looks pretty good. During 2002–6 Jeter's fielding percentage was below average only during 2003 (he was injured during much of the 2003 season). In contrast, Furcal had an above-average fielding percentage only during 2005. So this super-ficial analysis indicates that Jeter is a much better shortstop than Furcal. Not so fast! The problem with Fielding Percentage is that it does not take into account the balls a player does not get to; a player cannot make an er-ror on a ball he does not get to. If a SS does not move, he will field easy balls and make few errors. An immobile shortstop will allow many more base hits than a shortstop with great range.

The Range Factor: An Improved Measure of Fielding Effectiveness

How can we measure whether a shortstop has great range or poor range? Bill James developed an ingenious yet simple measure of fielding effec-tiveness, which he calls the Range Factor (RF). James defines a fielder's RF as the sum of putouts and assists a fielder gets per game played. Then James normalizes this statistic relative to all players in a given position. It turns out that shortstops during 2000–2006 average 4.483 PO + A per game. Thus a SS who had 5 PO + A per game would have an RF of $5/4.48 = 1.11$. This SS fields 11% more balls than a typical SS. Shortstops with an RF larger than 1 have above-average range and shortstops with range factors less than 1 have below-average range. Let's compute Derek Jeter's 2006 RF. We assume 8.9 innings per game. The column InnOuts indicates that in 2006 Jeter was the on field for $3,877/(8.9 \times 3) = 145.2$ games. Jeter had $PO + A = 214 + 381 = 595$. Thus Jeter successfully handled $\dfrac{595}{145.2} = 4.098$ chances per game. This is far below the average SS, who successfully handled 4.48 chances per game. Thus Jeter's normalized RF is $\dfrac{4.098}{4.483} = .91$. This implies that in 2006 Jeter successfully handled 9% fewer chances than an average SS. In contrast, Furcal in 2006 handled 10% more chances than an average SS. For the sake of comparison, Ozzie Smith (who played short-stop for the Padres and Cardinals from 1978 to 1996) is generally consid-ered the greatest fielding shortstop of all time. Ozzie had a lifetime fielding average of .978 (slightly above average), but he had several years where his RF exceeded 1.3. For Ozzie Smith, the Range Factor metric shows his true greatness.

Later in the chapter we will discuss how much the balls that Jeter does not field cost the Yankees.

Problems with Range Factor

There are several problems with RF. Suppose that SS1 plays for a team where pitchers strike out an average of 8 hitters per game and SS2 plays for a team whose pitchers strikeout only 5 batters a game. SS2's team will on average face three more balls in play than SS1's team, so even if both short-stops have equal ability, SS2 will have a higher RF. Suppose SS1's team has primarily left-handed pitchers and SS2's team has primarily right-handed pitchers. Then most managers will stack their lineups against SS1's team with right-handed batters to take advantage of the platoon effect (see chap-ter 12). Right-handed batters are believed to hit more ground balls to shortstop than are left-handed batters. SS2's opponents will use primarily left-handed batters (who are believed to hit to shortstop less often than are right-handed batters). In such a situation SS1 would have more balls hit near him and would tend to have a larger RF than would SS2.

When evaluating the RF of outfielders we must realize that the park dimensions have a significant effect on the number of opportunities an outfielder will have to field successfully. For example, in spacious Dodger Stadium the left fielder will be able to make many more putouts than the left fielder in Fenway Park (whose Green Monster prevents many fly balls from being caught). Suffice it to say that sabermetricians understand these problems and have created adjusted RFs to account for these and other problems.

The Fielding Bible: A Great Leap Forward

I believe that John Dewan (author of *The Fielding Bible*) has developed an outstanding way to evaluate fielding effectiveness. Dewan and his col-leagues at Baseball Info Solutions watch videotape of every MLB play and determine how hard each ball was hit and which "zone" of the field the ball was hit to. Then they determine the chance (based on all plays during a sea-son) that a ball hit at a particular speed to a zone would be successfully fielded. For example, they might find that 20% of all balls hit softly over second base are successfully fielded by shortstops. A shortstop who success-fully fields such a ball has prevented one hit. An average fielder would have successfully fielded this ball 20% of the time, so our SS has prevented $1 - .2 = .8$ hits more than an average player. In this case our SS receives a score of $+ 0.8$ on the batted ball. If our shortstop fails to make the play, he has prevented $0 - .2 = -.2$ hits and he receives a score of -0.2 on the batted ball. Note if our SS successfully fields 1 in 5 chances in this zone his net score is $.8 - 4(.2) = 0$, as we would hope. If over the course of a season a SS has a net score of -20, then he has effectively given up 20 more hits than has an average fielder. A SS with a score of $+30$ has effectively prevented 30 more hits than has an average fielder.

Converting Fielder's Scores to Runs

Can we convert a fielder's score to runs (and possibly games won or lost due to fielding)? To convert an extra hit allowed or saved to runs we must look at table 6.2, which gives expected runs in all possible states. Suppose a SS fails to field a ball with 0 outs and bases empty. Before this ball was hit the state was 0000 and an average team was expected to score 0.54 runs. If the SS gives up a hit to the next batter the new state is 0100 and the average batting team is expected to score 0.93 runs. If the SS turns a potential hit into an out, then the new state is 1000 and the average batting team is expected to score only 0.29 runs. Thus, in this situation the SS's failure to prevent a hit cost his team $0.93 - 0.29 = .64$ runs. If we average the cost of allowing an out to become a hit over all possible states (weighting each state by the fraction of the time each state occurs), we find that a hit allowed costs a team around 0.8 runs.

Converting Runs Saved by a Fielder into Wins

How do we convert runs saved by a good fielder or extra runs allowed by a bad fielder into wins? Let's look back at the Pythagorean Theorem of chapter 1. An average team scores 775 runs and gives up 775 runs (for the years 2000–2006) during a season. Of course, an average team wins 81 games. If a fielder saves 10 runs, then our average team now outscores its opponent 775–765, which yields a scoring ratio of $775/765 = 1.013$. Using the Pythagorean Theorem this translates into the team now winning $\dfrac{162(1.013^2)}{1.013^2 + 1} = 82.05$ games. Therefore, 10 runs translates into around 1 game won. This implies that a fielder whose *Fielding Bible* rating was -12.5 would cost his team around 1 game a year. Most front-line players have a *Fielding Bible* rating between $+20$ and -20. Therefore, few high-caliber fielders cost their team or save their team 2 more wins than average fielders do. In chapter 8 we will learn how to combine a batter's fielding and hitting abilities to obtain a measure of overall player effectiveness.

	B	C	D
3		**Team Total**	**−139**
4	**Position**	**Player**	**Fielding Bible rating**
5	1B	Giambi	−8
6	2B	Cano	−27
7	SS	Jeter	−34
8	3B	Rodriguez	2
9	LF	Matsui	3
10	CF	Williams	−37
11	RF	Sheffield	−38

Figure 7.3. Yankees' *Fielding Bible* ratings, 2005. See John Dewan, *The Fielding Bible* (Acta Sports, 2006).

Why Do the Yankees "Underperform"?

Most baseball fans are surprised that the Yankees, with their huge payroll, do not win the World Series every year. A major problem with the Yankees is their poor fielding.

Figure 7.3 gives the ratings of the Yankees' starting players for the 2005 season using Dewan's *Fielding Bible*. We see that the Yankee fielders cost the team 139 hits over the course of the season. This translates into $139 \times (.8) = 111.2$ runs. This means that the Yankees' fielding was $111.2/10 = 11.2$ wins worse than an average team's fielding.[1]

Derek Jeter vs. Adam Everett and Rafael Furcal

To close the chapter, let's compare Derek Jeter and Rafael Furcal to the major league's best shortstop, Adam Everett. Table 7.1 compares the *Fielding Bible* ratings for Jeter, Furcal, and the best fielding shortstop, Adam Everett of the Houston Astros, during the 2003–5 seasons.

The *Fielding Bible* ratings show (as does RF) that Furcal is a much better shortstop than Jeter, and Everett is by far the best of the three shortstops. Over the course of the 2003–5 seasons, Everett's +76 total compared to Jeter's −64 total implies that if Everett replaced Jeter as Yankee SS (and if he hit as well as Jeter), then our best estimate is that during the 2003–5 seasons the Yankees would have won 11.2 more games (or around 4 games per season).

During the 2003–7 seasons Adam Everett averaged a +29 *Fielding Bible* rating. This means that on average (when compared to an average fielding

---

[1] During their 2005 playoff series against the Tigers, the Yankees made five errors and were eliminated in four games.

TABLE 7.1
Fielding Bible Ratings for Derek Jeter, Rafael Furcal,
and Adam Everett

Year	Jeter Rating	Furcal Rating	Everett Rating
2003	−14	+10	+21
2004	−16	+2	+22
2005	−34	+26	+33

Source: John Dewan, *The Fielding Bible* (Acta Sports, 2006), 119–20.

shortstop), Everett's fielding generated $29(.8)/10 = 2.3$ wins per season for the Houston Astros.

SAFE: Spatial Aggregate Fielding Evaluation

Suppose two zones (say, 1 and 2) are adjacent and an average fielder is expected to field 20% of the balls hit to zone 1 and 40% of the balls hit to zone 2. What probability of being fielded should be used for a ball hit between the two zones? This is a difficult question to answer. Ideally the probability that an average fielder will field a ball should depend continuously on where the ball is hit and the speed at which the ball is hit. Several Wharton statisticians have performed such an analysis (called Spatial Aggregate Fielding Evaluation [SAFE]). The details of their methodology and their fielding ratings are posted on their Web site, http://stat.wharton.upenn.edu/~stjensen/research/safe.html.

8

PLAYER WIN AVERAGES

Anyone associated with a baseball, football, or basketball team would probably say that a player's objective is to help his team win games. Therefore, it seems reasonable to measure how much a professional athlete's efforts help his team win or cause his team to lose games. As we will see in later chapters, for basketball and football this is a very difficult task. For baseball, however, Eldon Mills and Harlan Mills (*Player Win Averages*) came up with a simple yet elegant way to measure how a baseball player changes the chance that his team will win a game. To illustrate the method, consider perhaps the most famous hit in baseball history: Bobby Thompson's home run in the 1951 playoffs. Thompson came to bat for the New York Giants in the bottom of the ninth inning of the deciding game of the 1951 playoff against the Brooklyn Dodgers. The Giants were down 4–2 and had runners on second and third with one out. Assuming the two teams were of equal ability, we can calculate that at this point (more on this later[1]) the Giants had a 30.1% chance of winning. Thompson hit his historic home run and the Giants won. Of course, the Giants now had a 100% chance of winning. So how can we measure the credit Thompson should be given for this batter-pitcher interaction? At the start of the game we assume each team has a 50% chance of winning. The metric we track at all points in time is

(my team's chance of winning) − (opponent's chance of winning).

Let's call this metric Winning Probability Difference (WINDIFF). At the start of a game, WINDIFF = 50 − 50 = 0. After each game event (batter outcome, stolen base, pickoff, and so forth) the batter and pitcher receive credit equal to how they change the value of WINDIFF.

---

[1] The most complete list of game-winning probabilities based on game margin, inning, out situation, and runners on base is in Tango, Lichtman, and Dolphin, *The Book*, chapter 1.

Before Thompson's home run the Giants had a 30.1% chance of winning (more later on how we compute this probability). Therefore WINDIFF = 30.1 − 69.9 = −39.8. After Thompson hit his home run, WINDIFF = 100 − 0 = 100. Therefore, Thompson's historic home run earns him 100 − (−39.8) = 139.8 WINDIFF points and costs the pitcher (poor Ralph Branca!) 139.8 WINDIFF points.

Jeff Sagarin's Player Win Averages

I will be using Jeff Sagarin's Player Win Average analysis for the 1957–2006 seasons in this section.[2] His widely respected baseball, basketball, and football statistical analyses have been published in *USA Today* for over twenty years.[3]

To avoid ugly decimals Sagarin multiplied WINDIFF scores by 10. Thus for Sagarin, before the home run WINDIFF = −398 and after the home run WINDIFF = 1,000, so Thompson receives 1,398 WINDIFF points for his home run. We will use SAGWINDIFF to denote Sagarin WINDIFF points for a given situation.

Here's another example. The leadoff hitter for the visiting team hits a double. How many SAGWINDIFF points does he earn? Before he hit the double SAGWINDIFF = 10(50 − 50) = 0. After he hits the double the chance of winning the game is 55.6% so SAGWINDIFF = 10(55.6 − 44.4) = 112 SAGWINDIFF points. In this case the hitter gains 112 SAGWINDIFF points and the pitcher loses 112 WINDIFF points.

Here is one final example. The home team is losing by two runs in the bottom of the ninth inning with the bases loaded and none out. Mariano Rivera comes in to pitch and the batter hits into a double play, scoring the runner from third. Before the double play the home team had a 52.3% chance of winning. After the double play the home team has a runner on third and two outs and is down one run. In this situation the home team has a 17.2% chance of winning. Thus, before the double play, SAGWINDIFF = 10 × (52.3 − 47.7) = 46. After the double play, SAGWINDIFF = 10 × (17.2 − 82.8) = −656. The batter gains −656 − 46 = −702 points (or loses 702 points), while Rivera gains 702 points. Notice Player

---

[2] Benjamin Polak and Brian Lonergan of Yale University also have a Player Win Average system. See "What's a Ball Player Worth," http://www.businessweek.com/bwdaily/dnflash/nov2003/nf2003115_2313_db016.htm.

[3] Sagarin's Player Win Average analysis is available at http://www.kiva.net/~jsagarin/mills/seasons.htm.

Win Averages resolves an important problem with ERA. The pitcher replaced by Rivera will have lost points for loading the bases (as he should), but Rivera gains, as he should, a lot of points for forcing the batter to hit into the double play. With ERA, the pitcher Rivera replaced suffers no penalty for loading the bases.

A Key Fact: 2,000 SAGWINDIFF Points = 1 Win

With SAGWINDIFF 2,000 points equals one game won. Once you understand this you can gain many insights. Too see how one win equals 2,000 points, let's suppose a game simply consisted of three events. Let's look at a game we win and a game we lose (see tables 8.1 and 8.2). The key thing to note from tables 8.1 and 8.2 is that in a win a team gains 1,000 points and in a loss a team loses 1,000 points. Let's consider a team that is one game over a 0% winning percentage. Their record is 82-80. They will have $82(1,000) - 80(1,000) = 2,000$ points for season. Therefore, total SAGWINDIFF points for the team = $2,000 \times$ (number of games over .500).

Let's look at the leaders in SAGWINDIFF for pitchers and hitters for the last three seasons (see table 8.3). The Situations column gives plate appearances for a hitter or batters faced for a pitcher. Recall that 2,000 points equals one win. Thus in 2006 Albert Pujols generated around nine more wins with his plate appearances than would have been generated by an average MLB player. Relief pitcher Francisco Rodriguez generated around five more wins with his pitching than would have been generated by an average MLB pitcher.

Incorporating *Fielding* Ratings
in Player Win Averages

Using John Dewan's *Fielding Bible* statistics we discussed in chapter 7, we can easily adjust a Player Win Average to include a player's fielding ability. Recall that in 2005 Derek Jeter had a *Fielding Bible* rating of -34 and Adam Everett had a rating of $+33$. Recall a hit is worth 0.8 runs. Thus, Jeter's fielding (relative to average fielder) cost the Yankees 27.2 runs = 2.72 wins = 5,440 Player Win Points. Everett's fielding helped the Astros' defense save 26.4 runs = 2.64 wins = 5,280 points. In 2005 Jeter had $+4,140$ Player Win Points and Everett had $-4,297$ Player Win Points. After adjusting their Player Win Points for fielding ability we find that Jeter had $4,140 - 5,440 = -1,300$ points in 2005 and Everett had $-4,297 + 5,280 = +983$ points in 2005. After adjusting for fielding

TABLE 8.1
SAGWINDIFF Analysis of a Win

Event	Our Win Chance	Opponent Win Chance	SAGWINDIFF Score
Start of game	50%	50%	
Event 1	20%	80%	−600
Event 2	40%	60%	+400
End of game	100%	0%	+1200

TABLE 8.2
SAGWINDIFF Analysis of a Loss

Event	Our Win Chance	Opponent Win Chance	SAGWINDIFF Score
Start of game	50%	50%	
Event 1	20%	80%	−600
Event 2	40%	60%	+400
End of game	0%	100%	−800

TABLE 8.3
Sagarin Win Average Leaders, 2004–6

Year	Player	Position	Total Points	Situations
2006	Albert Pujols	batter (outfield)	+18,950	653
2006	Francisco Rodriguez	relief pitcher	+10,562	312
2005	David Ortiz	outfielder/ designated hitter	+18,145	718
2005	Roger Clemens	starting pitcher	+12,590	852
2004	Barry Bonds	outfielder	+25,398	637
2004	Brad Lidge	relief pitcher	+11,906	382

ability, Everett helped the Astros more than Jeter helped the Yankees in 2005.

Recall that in chapter 7 we mentioned that the importance of fielding is overrated by many baseball analysts. Player Win Averages make this clear. Dewan's 2005 *Fielding Bible* ratings show that only seven players have a rating that would result in at least two wins more than would be generated by an average fielder. From Sagarin's 2005 Player Win Ratings we find that 44 batters and 46 pitchers have Player Win Ratings that generate two more wins than are generated by an average hitter or pitcher, respectively. This implies that there are many pitchers and hitters who have a large positive impact on team performance, but very few fielders have such an impact.

We can also adjust pitchers' Win Averages to account for a team's fielding quality. Recall from chapter 7 that the 2005 Yankees' fielding cost the team 11 games. This is equivalent to 22,000 Win Points. Therefore, we should increase the Win Point ratings of 2005 Yankee pitchers by 22,000 points. For example, Mariano Rivera faced 5% of all batters faced by Yankee pitchers in 2005. Therefore, we could add $.05(22,000) = 1,100$ Win Points to his total.

How Baserunning Ability Is Factored into Player Win Averages

A good base runner can certainly help his team win some games. Most good base runners help their team by

- stealing bases often and rarely getting caught when they try and steal;
- rarely grounding into double plays;
- taking the extra base when the batter gets a hit (for example, a fast base runner will score from second on a higher percentage of singles than will an average runner).

Player Win Averages reward a runner who steals a base and penalize a runner who is caught stealing. Therefore, Player Win Averages incorporate a player's base-stealing ability. Since grounding into a double play usually greatly reduces a team's chance of winning, Player Win Averages penalize a player who grounds into a double play. Therefore, Player Win Averages pick up a hitter's ability to avoid double plays.

A speedy base runner who often scores from second on a single or from first on a double, or often goes from first to third on a single is not rewarded by Player Win Averages. Dan Fox has analyzed how many runs a good base runner can add to a team by taking the extra base more often than an average base runner would add.[4] Fox found that during the 2000–2004 seasons Luis Castillo added the most runs (13.7, or 2.7 runs per season) while Edgar Martinez cost his team the most runs (12.6, or 2.5 runs per season). Since 10 runs equal one win, a runner who is great (or poor) at taking the extra base makes almost no difference in terms of costing his team runs (this equates to less than 600 SAGWIN Points, which is very small).

Did the 1969 Mets Win with Pitching or Hitting?

We can use Player Win Averages to determine how much of a team's success (or failure) can be attributed to the team's hitting and pitching. Let's look at Player Win Averages for the World Champion 1969 Mets. See figure 8.1.

	A	B	C	D	E
2		Hitters total		Pitchers total	Team total
3		−9319		41057	31738
4	Batter	Winpoints	Pitcher	Winpoints	
5	Grote	−1960	Seaver	13471	
6	Kranepool	−765	Koosman	13218	
7	Boswell	−42	Cardwell	761	
8	Garrett	−3819	McAndrew	1332	
9	Harrelson	−131	Ryan	23	
10	Agee	5410	McGraw	10902	
11	Jones	6334	Koonce	185	
12	Swoboda	6278	DiLauro	890	
13	Shamsky	4998	Taylor	1625	
14	Weis	−2054	Frisella	−605	
15	Gaspar	−1772	Jackson	−745	
16	Pfeil	−3629			
17	Clendenon	240			
18	Martin	−2196			
19	Charles	−1021			
20	Otis	−1789			
21	Dyer	−290			

Figure 8.1. Mets' Player Win Averages, 1969.

The Mets' hitters (including the pitcher) generated −9,319 points, indicating Mets batters performed worse than average batters would have (a loss of 4.6 games). On the other hand, the Mets' pitchers (including the Mets' defense) accumulated 41,057 points, which indicates the Mets'

[4] Dan Fox, "Circle the Wagons: Running the Bases Part III," August 11, 2005, http://www.hardballtimes.com/main/article/circle-the-wagons-running-the-bases-part-iii.

pitchers performed better than a staff of average pitchers would have (a gain of 20.5 games). This confirms the view that Tom Seaver, Jerry Koosman, and Tug McGraw were the key reason why the Mets performed so well during the 1969 regular season. The astute reader might observe that the Mets were 100-62, which should yield for the Mets $(100 - 81) \times 2{,}000 = 38{,}000$ Win Points. The Mets accumulated 31,738 points because several Mets were traded during season and Sagarin's totals are based on a player's record for the entire season. In chapter 12 we will see that the great clutch hitting of batters like Art Shamsky and Ron Swoboda was also an important factor in the 1969 Mets' Cinderella season.

In chapter 9 we will see how Player Win Averages can be used to evaluate trades and to determine a fair salary for a player.

Estimating Winning Probabilities
for Different Game Situations

Let's briefly discuss how to estimate the winning probability given various run margins, what inning a game is in, how many outs there are in a given inning, and the on-base situation in a given inning. One way of doing this is simply by using a large sample of major league games and looking at the percentage of the time a team wins. Christopher Shea's terrific Win Expectancy Finder Web site does just that.[5] Figure 8.2 uses the situation given earlier in which Bobby Thompson played a key role as an example of how to use the Web site. The Win Expectancy Finder gives the (based on the 1977–2006 seasons) percentage of the time the home team has won a game given the inning, number of outs, on-base situation, and current score differential.

During the years 1977–2006 the home team was down two runs and had runners on second and third and one out in the bottom of ninth a total of 203 times. The home team won $62/203 = 30.5\%$ of these games. Note that *The Book* estimated the chance of winning in this situation as 30.1%, which is very close to the probability implied by the actual game results over the last thirty seasons. *The Book* uses Markov chain analysis (which is equivalent to our chapter 4 simulation model) to estimate the probability the home team would win in all possible situations (there are thousands of possible situations).[6] By

---

[5] See http://winexp.walkoffbalk.com/expectancy/search.

[6] If we assume the run margin is always between -10 and $+10$ runs, then there are $24 \times 18 \times 21 = 9{,}072$ possible situations. I obtained this result by multiplying the number of states during an inning (24) by the number of possible half innings (9×2), and then multiplied by the 21 possible run differentials.

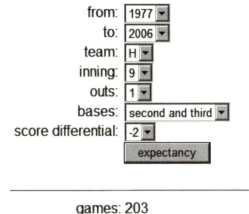

Figure 8.2. Example of Win Expectancy Finder.

using Markov chain analysis or our simulation model you can play out any situation thousands or even millions of times. For example, there are not many games in which the home team is down, say, twelve runs after one inning, so it is hard to accurately estimate the probability the home team will win in this situation. Using simulation or Markov chain analysis, we can play out this situation millions of times and obtain a more accurate estimate of the home team's winning probability.

Once we have an accurate estimate of a team's chance of winning in each possible run margin, inning, out, or base-runner situation, it is a simple to compute the Player Win Ratings. Simply find each event involving a player from box scores[7] and total his SAGWINDIFF points corresponding to each event in which the player is involved.

---

[7] These are available at http://www.retrosheet.com.

9

THE VALUE OF REPLACEMENT PLAYERS

Evaluating Trades and Fair Salary

In this chapter we will learn how to use the Player Win Averages discussed in chapter 8 to evaluate trade offers and calculate a player's fair salary (based on his previous year's performance).

VORPP: Value of a Replacement Player Points

The key tool involved in our analysis will be the Value of a Replacement Player Points (VORPP), which was developed by Keith Woolner, formerly of the *Baseball Prospectus* and now an executive for the Cleveland Indians. You may recall in several earlier chapters we compared a player's batting, pitching, and/or fielding performance to that of an average player. Although such comparisons are interesting, they do not really help us determine the true value of a player. We know that players create value by doing good things. Woolner appears to be the first person to have realized that players also create value by keeping bad players out of the lineup. Woolner asked what a team would do if a player were injured. The team often brings up from the minor leagues a player whose salary is very low (let's say 0). In theory there is a near inexhaustible supply of such players (called replacement players). To define a replacement player for, say, second basemen, Woolner would put the second basemen in descending order of plate appearances and define the replacement second basemen to be those who rank in the bottom 20% of this list. Woolner calculated that a lineup consisting totally of replacement players would generate a season record of 44–118 (winning four more games than the 1962 New York Mets).

Recall from chapter 8 that for each game a team finishes below .500 they will earn −2,000 SAGWIN points. Therefore, a team of replacement players should end the season with −74,000 points. Ignoring the relatively small impact of fielding,[1] it seems reasonable to assume that the replacement pitchers and batters should share equally the "blame" for a team's poor performance. This implies that a team of replacement batters would generate −37,000 points for the season as would a team of replacement pitchers. An average major league team has around 6,200 plate appearances during a season. This means per plate appearance a replacement hitter averages −37,000/6,200 = −5.97 points per plate appearance. Similarly, a replacement pitcher averages −5.97 points per batter faced (BFP). Now we see how an "average" hitter or pitcher has value. Consider an average hitter, Joe Hardy, with 500 plate appearances. Through his 500 plate appearances, however, Joe saved 5.97(500) = 2,985 points, or almost 1.5 games, A similar analysis applies to each pitcher. For each hitter we may now define VORPP as the following:

$$\text{VORPP} = \text{SAGWIN points} + 5.97 \times (\text{plate appearances}). \qquad (1)$$

For each pitcher we may define VORPP as

$$\text{SAGWIN points} + 5.97 \times (\text{BFP}).$$

For example, during 2006 Albert Pujols earned 18,950 points in 653 plate appearances. Thus, Pujols's VORPP = 18,950 + 653(5.97) = 22,848. Thus Pujols's 2006 performance earned the Cardinals around 11.4 more wins than a replacement player.

Using VORPP to Evaluate Trades

In 2006 San Diego Padre relief pitcher Trevor Hoffman earned 7,963 points in 255 BFP, and Padre starting pitcher Chris Young earned 6,117 points in 781 BFP. Their VORPP are as follows.

$$\text{Hoffman VORPP} = 7,963 + 5.97(255) = 9,485.$$

$$\text{Young VORPP} = 6,117 + 5.97(781) = 10,780.$$

Thus Hoffman and Yong each added around five more wins than a replacement pitcher would have added.

Let's assume that Pujols makes the same salary as Young and Hoffman combined. Based on our VORPP information, the Cardinals would be making a bad deal if they traded Pujols for Hoffman and Young. The Car-

---

[1] According to *The Fielding Bible*, in 2005 substitute fielders had almost the same average rating as did starting fielders.

dinals would be giving up over 11 wins for 10 wins. This assumes, of course, that the player's past performance is a perfect predictor of future performance. The beauty of this approach is that we have a single metric that allows us to compare the value of relief pitchers and starting pitchers to the value of batters.

Using VORPP to Determine a Fair Player Salary

We can also use VORPP to set a fair salary value for a player. In 2006 the average team payroll was $77 million. An average team goes 81–81. A "costless" team of replacement players would go 44–118. Therefore, 74,000 VORPP would bring our replacement team to a .500 level of play. This implies that $77 million "buys" 74,000 VORPP. Therefore, 1 VORPP is worth $77 million/74,000 = $1,040. We therefore estimate that the "value" generated during 2006 by these players is as follows:

Pujols 2006 value = 22,848 × 1,040 = $23.8 million.
Hoffman 2006 value = 9,485 × 1,040 = $9.9 million.
Young 2006 value = 10,780 × 1,040 = $11.1 million.

Is Alex Rodriguez (A-Rod) Overpaid?

In November 2007, Alex Rodriguez signed a new contract with the New York Yankees that will pay him approximately $275 million for ten years. Given A-Rod's past productivity, was this a good deal for the Yankees? We can use A-Rod's 2003–6 SAGWIN statistics and the fact that each VORPP is "worth" $1,040 to determine what a fair salary for A-Rod would have been (in 2006 dollars) for his batting performance during the 2003–7 seasons.

Figure 9.1 indicates that A-Rod generated around $14 million per year in value. In 2007 dollars his new contract is equivalent to $202 million

	C	D	E	F	G
		Plate	**SAGWIN**		**Fair Salary in 2006 Dollars**
5	**Year**	**Appearances**	**Points**	**VORP Points**	**(millions)**
6	2003	741	10593	15016.77	$ 15.62
7	2004	734	6575	10956.98	$ 11.40
8	2005	752	12521	17010.44	$ 17.69
9	2006	695	2205	6354.15	$ 6.61
10	2007	736	14193	18586.92	$ 19.33

Figure 9.1. What is A-Rod worth? See file Arod.xls.

(assuming salaries increase 5% per year). Therefore, for A-Rod to generate fair value over the next ten years he would need to continue to produce at the level of his phenomenal 2007 season for the 2008–16 seasons.

In our computation of a fair salary for A-Rod, we are assuming each win generated by a major league team is of equal value. As pointed out in Vince Gennaro's book *Diamond Dollars*, this is probably not true for two reasons.

- An extra win generates more revenue for a large market team like the Yankees than it would for a small market team like the Kansas City Royals.
- Even for two teams in a market of identical size, an extra win generates much more value for a team that has a good chance to make the playoffs than for a team that has a small chance of making the playoffs. This is because a playoff appearance generates $14 million in value to a major league team.[2] Therefore, A-Rod is of much more value to a team like the Yankees, which has a good chance of making the playoffs each year, than he is to a team that is either a playoff lock or has a small chance of making the playoffs.[3]

Extra Plate Appearances Create Value

A player who performs relatively poorly in many plate appearances may have a better VORPP than a player who performs very well in fewer plate appearances. For example, in 2006 Detroit Tiger third baseman Brandon Inge earned only 15 points in 617 plate appearances while Cubs shortstop Ryan Therriot generated a whopping 2,480 points in only 174 plate appearances. Per plate appearance, Inge generated roughly 0 points while Therriot generated 14 points per plate appearance. Despite this fact, Inge comes out better on VORPP because he replaces 617 potentially "bad" plate appearances.

$$\text{Inge VORPP} = 15 + 5.97(617) = 3,699.$$

$$\text{Therriot VORPP} = 2,480 + 5.97(174) = 3,519.$$

As we've seen, an average player creates value by keeping bad players from playing.

A good example of how VORPP works is presented in *Mind Game* by Steven Goldman, which details how the Boston Red Sox, led by savvy GM

---

[2] Baseball Prospectus Team of Experts, Keri, and Click, *Baseball between the Numbers*, 186.
[3] See Gennaro's *Diamond Dollars* for more details about how to incorporate pennant race likelihoods and market size into salary determination.

Theo Epstein, used VORPP to evaluate many of the player transactions that led to the Red Sox 2004 World Series title. For example, Goldman describes how Keith Foulke's high VORPP for the 1999–2003 seasons led the Red Sox to sign Foulke (a relief pitcher) for 2004. Foulke recorded 32 saves during the 2004 season. For example, in 2003 Foulke had a VORPP of 10,998, while the more highly touted Yankee reliever Mariano Rivera had a VORPP of only 9,021.

10

PARK FACTORS

During the 2006 season right fielder Brad Hawpe of the Colorado Rockies had a basic Runs Created rating of 5.04 runs per game. During the 2006 season San Diego Padre second baseman Josh Barfield had a basic Runs Created rating of 4.21 runs per game. On the surface, this would seem to indicate that Hawpe had a much better hitting season than did Barfield. Most baseball fans, however, realize that the Rockies play in Coors Field, which is notorious for being a hitter's park because the air is thin (the ball carries farther) and the park is not that big. On the other hand, the Padres play in spacious Petco Park, and the Padres are routinely involved in low-scoring games. Does the fact that Hawpe played in a hitter's park and Barfield in a pitcher's park mean that Barfield actually had a better hitting season than did Hawpe? As we will soon see, Barfield and Hawpe had virtually identical hitting seasons.

Bill James was the first to develop the concept of Park Factors. In every NBA arena the court is the same size and the baskets are ten feet high. In every NFL stadium the fields are the same dimensions (although Denver's thin air, domed stadiums and inclement weather may affect performance). In baseball, however, each stadium has different dimensions, which certainly influences how many runs are scored in the park. Park Factors are an attempt to measure how the park influences runs scored, home runs hit, and so forth.

We will discuss the simplest version of Park Factors. How much easier is it to score runs or hit a home run in Coors Field than it is in a typical National League park? Simply calculate runs scored per game in Coors Field divided by runs scored per road game. As shown in figure 10.1, during the 2006 season 10.73 runs per game were scored in Coors Park, and during road games, the Rockies scored 9.33 runs per game. In both road and home games, runs are equally affected by the Rockies' offense and defense, the average National League team's offense, and average National League

	B	C	D	E	F
2					**2006 Park Factors**
3	**Coors Field**	**Rockies**			
4		Home Runs	Runs	At Bats	Runs per game
5	Home	168	869	5653	10.72839506
6	Road	144	756	5509	9.333333333
7	Factor	1.136947933	1.1494709		
8					
9	**PETCO Park**	**Padres**			
10		Home Runs	Runs	At Bats	Runs per game
11	Home	167	652	5542	8.049382716
12	Road	170	758	5591	9.358024691
13	Factor	0.991038487	0.8601583		
14		Team	Raw Runs Created	Park Adjusted Runs Created	
15	Brad Hawpe	Rockies	5.04	4.6895262	
16	Josh Barfield	Padres	4.21	4.5264965	

Figure 10.1. Examples of Park Factors. See file Parkfactors.xls.

defense.[1] Therefore, the only difference between the runs scored in Coors Field and during away games must be a result of the influence of Coors Field. Thus $10.73/9.33 = 1.15$ times as many runs are scored in Coors Field as are scored in an average National League park.[2] We call this the Coors Field Park Factor. About 15% more runs are scored in Coors Field than in an average National League park. Similarly, we find that the Park Factor for the Padres' Petco Park is $8.05/9.35 = .86$. This indicates that in Petco Field, 14% fewer runs are scored than in an average National League park.

Now we can adjust Hawpe's and Barfield's Runs Created per game to account for the difference in Park Factors. In half of Hawpe's games he had a 15% advantage in scoring runs. Overall, this means we should deflate his Runs Created by dividing his actual Runs Created: $(1 + 1.15)/2 = 1.075$. Similarly, Barfield had a handicap of 14% in half his games so we should inflate his actual Runs Created: $(1 + .86)/2 = .93$. Now we find the following Adjusted Runs Created for each player.

Hawpe adjusted runs created $= 5.04/1.075 = 4.69$ runs created per game

[1] I am ignoring interleague play in this analysis,

[2] Of course, this ignores the fact that Coors Field is part of the National League average, but Coors Field is not included in any Rockies road game. This fact has a small effect on the Park Factor and we will ignore this problem.

Barfield adjusted runs created = 4.21/.93 = 4.53 runs created per game

Thus, after adjusting Barfield's and Hawpe's Runs Created per game to account for park effects, we find that their offensive performances were virtually identical.

In a similar fashion we can compute a home run adjustment for each park:

$$\text{HR adjustment for park} = \frac{\text{HR per AB in team's home games}}{\text{HR per AB in team's away games}}.$$

Thus for Coors Field the HR adjustment factor is $\dfrac{168/5653}{144/5509} = 1.14$. For Petco Park the HR adjustment factor is .99. This indicates that it is not the lack of home runs that caused fewer runs to be scored in Petco. We find, however, that there were 23% fewer doubles in Petco than during Padre road games. This indicates that it is harder to get an extra base hit in Petco than in the average National League park.[3]

Of course, we can adjust a pitcher's DIPS forecast for ERA to account for his Park Factor. For example, we should raise a San Diego Padre pitcher's DIPS by 1/.93, or around 8%. Recognizing that around 10 runs = 1 win = 2,000 SAGWINDIFF points, we can then adjust a hitter or pitcher's SAG-WINDIFF rating based on his Park Factor. For example, the park adjustment indicates that Barfield actually created 0.32 more runs per game than indicated by his basic Runs Created rating. Barfield caused 14.37 games' worth of outs, so he really created $(14.37) \times .32 = 4.60$ more runs that our basic rating indicates. This is worth 0.460 wins, or 920 points. Barfield's initial Player Win Rating was −606 points. Therefore, we should add 920 points to Barfield's 2006 Player Win Rating, leading to a final rating of −606 + 920 = 314 points.

---

[3] In Boston's Fenway Park, 37% more doubles are hit than in an average American League park.

11

STREAKINESS IN SPORTS

We have all heard Marv Albert tell us that Dirk Nowitzki is "on fire" or Jack Buck tell us that Albert Pujols is "red hot" and nobody can get him out. We also hear announcers tell us the Spurs are on a hot streak, are unbeatable, and so forth. Is it true that athletes and teams encounter hot streaks, or are the observed patterns of player and team performance just randomness at work?

What Does a Random Sequence Look Like?

Let's first examine how a random sequence of 162 wins and losses appears. Let's suppose a team wins 60% of their games. To generate a random sequence of 162 games, we should make sure that in each game the team has a 0.60 chance of winning and that the chance of winning a game does not depend on the team's recent history. For example, whether the team lost their last five games or won their last five games, their chance of winning the next game should still be 0.60. Figure 11.1 shows three randomly generated sequences of 162 games. In examining these data, most people would think they are looking at a "streaky team," even though these sequences were generated by assumptions that involve no streakiness.

These sequences were generated by essentially flipping a coin 162 times with a 0.60 chance of a win. A 1 denotes a win and a 0 a loss. First note that on average we would expect $162(.6) = 97.2$ wins and in none of our random sequences did this number of wins occur. This is because of the randomness inherent in the coin-tossing process. Also note that in each sequence the team experiences several long winning streaks. For example, in the second sequence the team had winning streaks of ten, nine, seven, and six games. Most people think the occurrence of winning streaks indicates momentum or a "hot team" effect, but here we see long winning streaks are simply random.

	Two 7 game streaks		One 5 game streak		
	Random Sequence 1				
Game	Outcome	Game	Outcome	Game	Outcome
1	1	55	0	109	1
2	0	56	1	110	0
3	1	57	0	111	1
4	1	58	0	112	0
5	1	59	0	113	0
6	1	60	1	114	0
7	0	61	0	115	1
8	0	62	0	116	1
9	1	63	1	117	1
10	1	64	0	118	1
11	0	65	0	119	1
12	0	66	1	120	1
13	1	67	0	121	1
14	1	68	1	122	0
15	0	69	1	123	1
16	1	70	1	124	1
17	0	71	0	125	1
18	1	72	1	126	1
19	0	73	0	127	0
20	1	74	1	128	0
21	0	75	0	129	0
22	0	76	0	130	1
23	0	77	0	131	1
24	1	78	1	132	0
25	0	79	1	133	0
26	0	80	0	134	0
27	0	81	0	135	1
28	1	82	1	136	0
29	0	83	0	137	1
30	1	84	0	138	0
31	1	85	1	139	1
32	0	86	1	140	0
33	1	87	1	141	0
34	1	88	0	142	1
35	1	89	1	143	1
36	1	90	0	144	1
37	1	91	1	145	1
38	1	92	0	146	1
39	0	93	0	147	1
40	1	94	0	148	1
41	1	95	0	149	0
42	0	96	0	150	1
43	1	97	1	151	1
44	1	98	1	152	0
45	1	99	0	153	1
46	0	100	0	154	1
47	0	101	1	155	1
48	1	102	0	156	1
49	1	103	1	157	1
50	1	104	1	158	0
51	1	105	1	159	0
52	1	106	1	160	1
53	1	107	1	161	0
54	0	108	0	162	0

Figure 11.1. Examples of random sequences.

	H	I	J	K	L	M
1	10, 8, 7, and 6 game streaks					
2		Random Sequence 2				
3	Game	Outcome	Game	Outcome	Game	Outcome
4	1	0	55	0	109	1
5	2	0	56	1	110	1
6	3	1	57	0	111	0
7	4	0	58	0	112	1
8	5	0	59	1	113	0
9	6	0	60	0	114	0
10	7	0	61	0	115	1
11	8	1	62	0	116	1
12	9	1	63	0	117	1
13	10	0	64	1	118	1
14	11	1	65	0	119	1
15	12	0	66	1	120	1
16	13	0	67	0	121	1
17	14	0	68	0	122	1
18	15	0	69	1	123	1
19	16	0	70	0	124	1
20	17	1	71	1	125	0
21	18	1	72	1	126	0
22	19	0	73	1	127	0
23	20	0	74	1	128	0
24	21	0	75	1	129	1
25	22	0	76	0	130	1
26	23	1	77	0	131	1
27	24	1	78	0	132	0
28	25	1	79	1	133	0
29	26	0	80	1	134	0
30	27	1	81	0	135	0
31	28	1	82	1	136	0
32	29	1	83	0	137	1
33	30	0	84	0	138	1
34	31	1	85	0	139	1
35	32	0	86	0	140	1
36	33	0	87	1	141	0
37	34	1	88	1	142	0
38	35	1	89	1	143	1
39	36	0	90	1	144	1
40	37	0	91	1	145	1
41	38	1	92	1	146	0
42	39	0	93	1	147	1
43	40	1	94	0	148	1
44	41	1	95	1	149	1
45	42	1	96	0	150	1
46	43	1	97	1	151	1
47	44	0	98	1	152	1
48	45	1	99	1	153	1
49	46	1	100	1	154	1
50	47	1	101	1	155	1
51	48	1	102	1	156	0
52	49	0	103	0	157	0
53	50	0	104	1	158	1
54	51	0	105	1	159	1
55	52	1	106	0	160	0
56	53	1	107	1	161	0
57	54	0	108	0	162	1

Figure 11.1. (cont.)

	A	B	C	D	E	F
		Random Sequence 3		Two 5, 6, 8, and 9 game streak		
58						
59	Game	Outcome	Game	Outcome	Game	Outcome
60	1	1	55	0	109	1
61	2	1	56	0	110	1
62	3	1	57	1	111	1
63	4	1	58	0	112	1
64	5	1	59	1	113	1
65	6	1	60	1	114	0
66	7	1	61	1	115	1
67	8	1	62	0	116	1
68	9	1	63	1	117	1
69	10	0	64	1	118	1
70	11	1	65	0	119	0
71	12	0	66	0	120	1
72	13	0	67	1	121	1
73	14	1	68	1	122	1
74	15	1	69	0	123	0
75	16	1	70	0	124	1
76	17	0	71	0	125	1
77	18	0	72	1	126	0
78	19	1	73	1	127	0
79	20	1	74	1	128	1
80	21	0	75	0	129	1
81	22	1	76	1	130	1
82	23	1	77	1	131	0
83	24	1	78	1	132	0
84	25	0	79	1	133	1
85	26	1	80	1	134	1
86	27	1	81	1	135	1
87	28	1	82	1	136	0
88	29	1	83	1	137	1
89	30	1	84	0	138	0
90	31	1	85	1	139	1
91	32	1	86	0	140	1
92	33	1	87	0	141	1
93	34	1	88	1	142	1
94	35	0	89	1	143	1
95	36	0	90	1	144	0
96	37	0	91	1	145	1
97	38	0	92	0	146	0
98	39	1	93	0	147	1
99	40	0	94	1	148	1
100	41	1	95	1	149	0
101	42	0	96	1	150	1
102	43	1	97	0	151	1
103	44	1	98	0	152	0
104	45	0	99	1	153	1
105	46	1	100	1	154	0
106	47	1	101	1	155	1
107	48	0	102	0	156	0
108	49	1	103	1	157	0
109	50	1	104	0	158	1
110	51	1	105	1	159	1
111	52	1	106	1	160	1
112	53	0	107	1	161	1
113	54	0	108	1	162	1

Figure 11.1. (*cont.*)

	F	G	H
			Probability of longest streak >
12	Longest streak	Probability	
13	1	0.0%	100.0%
14	2	0.0%	100.0%
15	3	0.0%	100.0%
16	4	0.2%	99.8%
17	5	3.0%	96.8%
18	6	11.5%	85.3%
19	7	17.4%	67.9%
20	8	19.7%	48.2%
21	9	16.5%	31.7%
22	10	11.5%	20.2%
23	11	7.8%	12.4%
24	12	4.7%	7.7%
25	13	2.9%	4.7%
26	14	2.0%	2.7%
27	15	1.0%	1.8%
28	16	0.7%	1.1%
29	17	0.4%	0.7%
30	18	0.2%	0.5%
31	19	0.1%	0.4%
32	20	0.1%	0.2%
33	21	0.2%	0.1%
34	22	0.0%	0.0%
35	23	0.0%	0.0%
36	24	0.0%	0.0%
37	25	0.0%	0.0%
38	26	0.0%	0.0%

Figure 11.2. Distribution of longest winning streaks in a random sequence.

Suppose a team has a 0.60 chance of winning each game and past performance has no influence on the chance of winning a game. Figure 11.2 shows the distribution of a team's longest winning streak for a 162-game season. For example, there is a 20% chance that a team with a .60 probability of winning each game will have a winning streak longer than ten games during the season.

Does the Hot Hand Exist?

Most people believe in the "hot hand" or streak hitter. In support of this view Gilovich, Vallone, and Tversky[1] found that 91% of all surveyed bas-

---

[1] Gilovich, Vallone, and Tversky, "The Hot Hand in Basketball."

ketball fans felt that basketball players were more likely to make a shot if the last shot were good than if the last shot were missed.

We can now look at whether the evidence supports the existence of the hot shooter, streak hitter, or streaky team. The most common technique used to test for streakiness is the Wald-Wolfowitz Runs Test (WWRT). Before discussing WWRT, however, let's quickly review some concepts of basic probability and statistics.

Enter the Normal Random Variable

A quantity such as the batting average of a randomly selected player or the IQ of a randomly selected person is uncertain. We refer to an uncertain quantity as a random variable. Given the value of a random variable such as a player's batting average or a person's IQ, can we determine how unusual the observation is? For example, which is more unusual: a 1980 major league batter who hits .360 or a person with an IQ of 140? To determine whether an observation is "unusual," we generally assume the data come from a normal distribution. For example, IQs are known to follow a normal distribution that is illustrated by the probability density function (pdf) shown in figure 11.3.

The height of the IQ probability density function for an IQ value x is proportional to the likelihood that a randomly chosen person's IQ assumes a value near x. For example, the height of the pdf at 82 is approximately half of the height of the pdf at 100. This indicates that roughly half as many people have IQs near 82 as near 100. A normal density is characterized by two numbers.

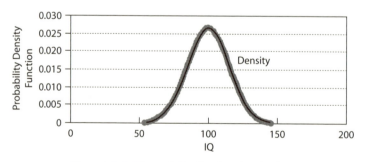

Figure 11.3. Normal distribution density for IQ.

- the mean μ or average value
- the standard deviation σ, which measures the spread of a random variable about its mean.

IQs are known to have μ = 100 and σ = 15. If a random variable follows a normal pdf, then

- the mean is the most likely value of the random variable;
- 68% of the time the random variable will assume a value within σ of its mean (the area under a pdf represents probability, so the total area under the pdf equals 1, and the area under the pdf between 85 and 115 for IQs is 68%);
- 95% of the time the random variable will assume a value within 2σ of its mean (thus 95% of all people have IQs between 70 and 130 and the area under the IQ pdf between 70 and 130 is .95);
- the pdf is symmetric (that is, it is just as likely that a normal random variable will assume a value near μ + x as near μ − x); thus roughly the same number of people have IQs near 120 as near 80, and so forth.

Z Scores

Assuming that a histogram or bar graph of our data tells us that the symmetry assumption fits our data, then statisticians measure how unusual a data point is by looking at how many standard deviations above or below average the data point is. This is called a z score. Thus

$$z \text{ score} = \frac{\text{data point} - \text{mean}}{\text{standard deviation}}.$$

When averaged over all data points, z scores have an average of 0 and a standard deviation of 1. This is why computing a z score is often called standardizing an observation. A z score of 2 in absolute value indicates that an observation more extreme than our data point has roughly a 5% chance of occurring. A z score of 3 in absolute value indicates that an observation more extreme than our data point has roughly 3 chances in 1,000 of occurring. As an example of the use of z scores, we might ask what was more unusual: the stock market dropping 22% on October 19, 1987, or seeing a person shorter than thirty inches tall walk down the street. We are given the following information:

mean daily stock return = 0%
sigma of daily stock returns = 1.5%

mean height of American male = 69 inches

sigma of height of American male = 4 inches

Then the z score for the stock market example $= \dfrac{-22-0}{1.5} = -14.67$,

while the z score for the short person example is $\dfrac{30-69}{4} = -9.75$.

Therefore a 22% market drop in a day is much more unusual than seeing a thirty-inch-tall person walk down the street.

Let's look at Rogers Hornsby's .424 batting average in 1924 and George Brett's .390 batting average in 1980 using z scores. Given that the pitching and fielding in both years was of equal quality, which player had the more outstanding performance? Figure 11.4 shows that the symmetry assumption is reasonable for batting averages.

For 1980, mean batting average = .274 and the standard deviation of batting averages = .0286.[2] Thus, George Brett's z score $= \dfrac{.390-.274}{.0286} = 4.06$. In 1924 mean batting average = .299 and standard deviation of batting averages = .0334. Therefore, Hornsby's z score $= \dfrac{.424-.299}{.0334} = 3.74$. Thus, even though Brett's batting average was 34 points lower than Hornsby's, relative to the overall performance during their respective seasons, Brett's performance was more outstanding.

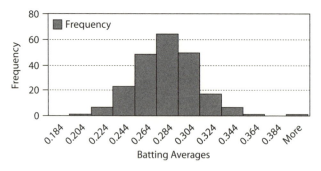

Figure 11.4. Histogram of 1980 batting averages.

[2] This takes into account only players who had at least 300 AB during the season.

The Wald Wolfowitz Runs Test (WWRT)

We now return to our discussion of the WWRT. To motivate the WWRT consider a team that is 5–5. The "streakiest" way to go 5–5 would be to have a sequence of wins and losses as follows: WWWWWLLLLL. The "least streaky" way to go 5–5 would be to have the sequence be WLWLWLWLWL. Define a "run" to be an uninterrupted sequence of Ws or Ls. Thus WWWWWLLLLL has two runs (WWWWW and LLLLL) while the sequence WLWLWLWLWL has 10 runs. This example makes it clear that few runs in a sequence is an indication of streakiness while many runs is an indication of a lack of streakiness. The key question is, given a random sequence (a sequence in which whether an observation is W or L does not depend on the prior events in the sequence), how many runs can we expect and how much spread is there about this average number of runs? Wald and Wolfowitz determined that for a random sequence consisting of successes (S) and failures (F), the mean and standard deviation of the number of runs are as follows. In these formulas, $N = S + F$ is the total sequence length.

$$\mu = \text{mean number of runs} = \frac{2FS}{N} + 1.$$

$$\sigma = \text{standard deviation of number of runs} = \sqrt{\frac{(\mu - 1)(\mu - 2)}{N - 1}}.$$

For example, if a team goes 5–5 on average, we would expect to see (if Ws and Ls are randomly sequenced) $\frac{2(5)(5)}{10} + 1 = 6$ runs with a standard deviation of $\sqrt{\frac{(6-1)(6-2)}{10-1}} = 1.49$. Therefore, the sequence WWWWWLLLLL has a z score of $\frac{2-6}{1.49} = -2.68$ while the sequence WLWLWLWLWL has a z score of $\frac{10-6}{1.49} = 2.68$.

An Introduction to Hypothesis Testing

Statisticians usually set up null and alternative hypotheses. The null hypothesis is to be accepted unless we observe a great deal of evidence in favor of the alternative hypothesis (similar to the U.S. justice system, in that the null hypothesis is "innocent" until proven guilty).[3] What is a great deal of evidence against the null hypothesis? Most statisticians believe that if the data indicate (under the assumption that the null hypothesis is true) that a result at least as extreme as what we have observed has less than a 5% chance of occurring, then we should reject the null hypothesis. Recall that if a random variable follows a normal distribution, there is only a 5% chance that the random variable's z score will exceed 2 (1.96, to be exact) in absolute value. Therefore, if we define a test statistic based on our data (given the null hypothesis is true) that follows a normal distribution, then there is only a 5% chance that the test statistic's z score will exceed 2 in absolute value. Therefore, if the test statistic's z score exceeds 2 in absolute value we should reject the null hypothesis. In this case we say the data are significant at the .05 level, because if the null hypothesis is true, then the chance of seeing a test statistic at least as extreme as what we observed is less than .05.

In our situation

null hypothesis: the Ws and Ls are randomly distributed within the sequence. That is, the past history of the sequence does not influence the chance that the next event is a W or L.

alternative hypothesis: the past history of the sequence has some effect on whether the next event is a W or L.

When trying to determine whether a sequence of Ws and Ls is random the appropriate test statistic is the number of runs. Suppose we observed WWWWWLLLLL or WLWLWLWLWL. In either case each z score exceeds 2 in absolute value, so either sequence would indicate that there is less than a 5% chance that the team's wins and losses came from a random sequence. We would reject the null hypothesis and conclude that the sequence is not random.

Back to the Hot Hand

We can now discuss Gilovich, Vallone, and Tversky's (GVT for short) analysis of whether basketball players' shooting exhibits the "hot hand" or "streak shooting." GVT analyzed for each home game of the 1980–81 Philadelphia 76ers season the results of successive field goal (FG) attempts. For

[3] In many court cases (such as *Castaneda vs. Partida* [1977]) the U.S. Supreme Court has accepted the 5% level of significance or two standard deviation rule as the level of evidence needed to shift the burden of proof from plaintiff to defendant or vice versa.

TABLE 11.1
Determining Whether the Hot Hand Existed for the 1980–81 Philadelphia 76ers

Player	Good FGs	Missed FGs	Actual Number of Runs	Expected Number of Runs	Z Score
Chris Richardson	124	124	128	125	−0.38
Julius Erving	459	425	431	442.4	0.76
Lionel Hollins	194	225	203	209.4	0.62
Mo Cheeks	189	150	172	168.3	−0.41
Caldwell Jones	129	143	134	136.6	0.32
Andrew Toney	208	243	245	225.1	−1.88
Bobby Jones	233	200	227	216.2	−1.04
Steve Mix	181	170	176	176.3	0.04
Darryl Dawkins	250	153	220	190.8	−3.09

example, GGGMMG would mean the player made his first three shots, missed his next two shots, and made his sixth shot. They performed a WWRT on each player's sequence of FG attempts. The results are shown in table 11.1.

Since only Darryl Dawkins has a z score for runs exceeding 2 in absolute value, he is the only player who exhibited significant streakiness. Perhaps the fact that most of Dawkins's baskets were dunks (his nickname was "Chocolate Thunder") and that he made his last shot indicate that he was being guarded by a player he could overpower. This would indicate that the next shot would be more likely to be a success.

How can we look at streakiness aggregated over all the 76ers? We simply average their z scores. Then, using the result that the standard deviation of the average of N independent z scores[4] is $1/\sqrt{N}$, we find that the average z

---

[4] A set of random variables is independent if knowledge of the value of any subset of the random variables tells us nothing about the distribution of the values of the other random variables. In our case, knowing the z scores of any subset of 76er players clearly tells us nothing about whether the other players are more or less likely to exhibit hot hand shooting behavior, so it is appropriate to assume that the z scores of individual players are independent random variables.

score for the nine players is -0.56. The standard deviation of the average of the nine z scores is $9^{-.5} = 1/3 = .333$. Thus the z score for "the average of z scores" is $\dfrac{-0.56 - 0}{.333} = -1.68$, which is not statistically significant at the .05 level. Thus we conclude that as a whole, the 76ers do not exhibit significant streakiness or hot hand shooting behavior.

Does Streak Hitting Exist in Baseball?

S. C. Albright[5] has analyzed the streak hitting of MLB players. He looked at players' sequences of hits and outs during a season. He computed the expected number of runs of hits (H) or outs (O) for each player and compared it to the actual number of runs (for example, sequence HHOOHO has four runs). Then he computed a z score based on the actual number of runs. Figure 11.5 uses Cal Ripken's player data from 1987 to illustrate Albright's method.

Ripken's positive z score indicates more runs occurred than were expected by chance. Therefore, in 1987 Ripken exhibited behavior that was slightly less streaky than expected. For all 501 players during the 1987 season, the average z score was -0.256. The standard deviation of the average of 501 z scores is $501^{-.5} = .045$. Thus the z score for the average of all 1987 z scores is $\dfrac{-.256 - 0}{.045} = -5.68$, which is significant evidence that players are streakier than average. Albright then used a more sophisticated technique,

	H	I	J
2	Runs	Hits	Outs
3	233	150	446
4	Mean	225.497	
5	Sigma	9.18294	
6			
7	z score		
8	0.8171		
9	p value		
10	0.41387		

Figure 11.5. Cal Ripken is not a streak hitter. See file 87ripc1.xls.

[5] Albright, "A Statistical Analysis of Hitting Streaks in Baseball."

logistic regression, to predict the probability that a player would get a hit on an AB based on his recent history as well as other variables including the following:

- Is the pitcher left- or right-handed?
- What is the pitcher's ERA?
- Is the game home or away?
- Is the game on grass or artificial turf?

After adjusting for these variables, the evidence of streakiness disappears. Albright also found that the players who exhibited significant streak hitting behavior during a given season were no more likely than a randomly chosen player to be a streak hitter during the next season. This indicates that streak hitting behavior does not persist from year to year.

"Hot Teams"

Do teams exhibit momentum? I analyzed the 2002–3 NBA season to determine whether teams exhibit streaky behavior. We cannot simply look at the sequence of wins and losses. Often during the NBA season a team will play six straight road games against tough teams, which can result in a long losing streak. The losing streak would in all likelihood be due to the strength of the opposition, not a variation in the team's performance. Therefore, for each game I created a "point spread" by including the strength of the team, the strength of opponent, and the home court edge. To illustrate the idea, suppose the Sacramento Kings are playing the Chicago Bulls in Chicago. The Kings played 7.8 points better than average during the 2002–3 season and the Bulls played 8.5 points worse than average. The NBA home edge is around 3 points. Since the Kings are 16.3 points ($7.8 - (-8.5) = 16.3$) better than the Bulls and the Bulls get 3 points for having the home edge, we would predict the Kings to win by 13.3 points. Therefore, if the Kings win by more than 13 points they receive a W and if they win by less than 13 or lose they receive an L. In short, a W indicates that a team played better than their usual level of play while an L indicates that a team played worse than their usual level of play. Figure 11.6 shows the results of using the WWRT to analyze the streakiness of the sequence of Ws and Ls for each team. We see that only Portland exhibited significant streakiness. Note, however, that with 29 teams we would (even if there were no streakiness) expect, by chance, $.05(29) = 1.45$ teams on average to have z scores exceeding 2 in absolute value. The average of the 29 team z scores is $-.197$. The standard deviation

	D	E	F	G	H	I	J
14	**Team**	**z score**	**Mean runs**	**Sigma**	**Actual runs**	**Wins**	**Losses**
15	Atlanta	0.888957	42	4.499657	46	41	41
16	Boston	0.45017	41.97561	4.496947	44	40	42
17	Charlotte	1.562035	41.97561	4.496947	49	42	40
18	Chicago	0.467286	41.902439	4.488816	44	43	39
19	Cleveland	0.672543	41.97561	4.496947	45	40	42
20	Dallas	−1.02608	40.439024	4.326202	36	49	33
21	Denver	0.005424	41.97561	4.496947	42	42	40
22	Detroit	−0.39785	41.780488	4.475265	40	44	38
23	Golden State	0.222239	42	4.499657	43	41	41
24	Houston	0.27368	40.804878	4.366855	42	48	34
25	Indiana	−1.55567	42	4.499657	35	41	41
26	L. A. Clippers	0.044682	40.804878	4.366855	41	34	48
27	L. A. Lakers	−1.10644	41.97561	4.496947	37	40	42
28	Memphis	−0.66672	42	4.499657	39	41	41
29	Miami	−0.64659	41.902439	4.488816	39	43	39
30	Milwaukee	0.467286	41.902439	4.488816	44	39	43
31	Minnesota	−1.03444	41.609756	4.456293	37	37	45
32	New Jersey	−0.39785	41.780488	4.475265	40	44	38
33	New York	0.888957	42	4.499657	46	41	41
34	Orlando	0.495951	41.780488	4.475265	44	38	44
35	Philadephia	−0.20104	41.902439	4.488816	41	39	43
36	Phoenix	−0.31369	41.390244	4.431901	40	36	46
37	Portland	−2.21831	41.97561	4.496947	32	42	40
38	Sacramento	−0.1744	41.780488	4.475265	41	44	38
39	San Antonio	−0.13683	41.609756	4.456293	41	45	37
40	Seattle	−0.21695	41.97561	4.496947	41	42	40
41	Toronto	−0.39785	41.780488	4.475265	40	38	44
42	Utah	−0.13683	41.609756	4.456293	41	37	45
43	Washington	−1.5151	41.780488	4.475265	35	38	44

Figure 11.6. Momentum for NBA teams. See file Teammomentum.xls.

of the average of 29 z scores is $\dfrac{1}{\sqrt{29}} = .186$, so the z score for the average

of all 29 team z scores is $\dfrac{-.197 - 0}{.186} = -1.05$, which is not significant

at the .05 level. Therefore, we conclude that the variation in team performance during the 2002–3 NBA season is well explained by random variation. This small study gives no support to the view that teams have
momentum or encounter more hot streaks than would be indicated in a
random sequence.

12

THE PLATOON EFFECT

For most right-handed pitchers, the curve ball is an important part of their pitching repertoire. A right-handed pitcher's curve ball curves in toward a left-handed batter and away from a right-handed batter. In theory, when facing a right-handed pitcher, a left-handed batter has an edge over a right-handed batter. Similarly, when a left-handed pitcher is on the mound, the right-handed batter appears to have the edge. Managers take advantage of this alleged result by platooning batters. That is, managers tend to start right-handed batters more often against left-handed pitchers and start left-handed batters more often against right-handed pitchers. Ignoring switch hitters, Joseph Adler found that 29% of batters who face left-handed pitchers are left-handed and 51% of batters who face right-handed pitchers are left-handed.[1] This shows that platooning does indeed exist. As the great American statistician and quality guru W. Edwards Deming said, "In God we trust; all others must bring data." Does actual game data confirm that a batter has an edge over a pitcher who throws with a different hand than he hits with? Adler tabulated OBP for each possible pitcher batter "hand" combination for the 2000–2004 seasons. The results are shown in figure 12.1, which tells us the following:

- Left-handed batters on average have a 22-point higher OBP against right-handed pitchers than against left-handed pitchers.
- Right-handed batters on average have a 13-point higher OBP against left-handed pitchers than against right-handed pitchers.
- Left-handed pitchers on average yield a 12-point larger OBP to right-handed batters than to left-handed batters.
- Right-handed pitchers on average yield a 23-point larger OBP to left-handed batters than to right-handed batters.

[1] Adler, *Baseball Hacks*, 330.

	C	D	E	F
3	Batter	Pitcher	Plate Appearances	OBP
4	L	L	35258	0.327
5	L	R	169180	0.349
6	R	L	86645	0.339
7	R	R	163214	0.326

Figure 12.1. Platooning results. See file Platoon.xls. Source: Joseph Adler, *Baseball Hacks* (O'Reilly Media, 2006), 331.

Most sabermetricians refer to these differences as platoon splits. Figure 12.1 shows that platoon splits definitely exist.

Since most major league pitchers are right-handed (76%), a left-handed hitter has an edge over a right-handed hitter because he will be facing mostly right-handed pitchers. This helps explain the seemingly amazing fact that 37% of major league hitters bat left-handed, when only about 10% of the U.S. population is left-handed.

Predicting a Player's Future Platoon Splits

During the 2005–7 seasons Jim Thome (a left-handed hitter) had a platoon split of 0.121, or 121 OBP points. Is it reasonable to expect that in 2008 he would have a 121-point platoon split when the average left-handed hitter has a 22-point platoon split? According to the theory of regression toward the mean, we would expect Thome's future platoon splits to be closer to average than were his past splits. The idea of regression toward the mean was first credited to Francis Galton in 1886. Galton observed that tall parents had parents who were taller than average but were closer to average height than their parents. Similarly, parents with shorter than average children tend to have children who are taller than they are but are still below average. Regression toward the mean explains many facts in sports. Consider the following:

- Highly successful NFL teams tend to be less successful the next year. See chapter 18.
- The famous *Madden NFL* and *Sports Illustrated* cover jinxes: If a player is on the cover of *Madden NFL* or *Sports Illustrated* he must have done something extraordinary. Therefore, regression toward the mean predicts the player will still do well but not as well as he did in the past.

In *The Book: Playing the Percentages in Baseball*, Tango, Lichtman, and Dolphin (hereafter TLD) found that the best way to predict a player's future platoon splits was to take a weighted average of the player's past platoon splits and the major league average. Their rules are summarized in table 12.1.

Note that the more data we have, the closer our prediction of a player's future split will be to his past platoon split. Also note that more weight is given to past platoon splits for pitchers than for batters.

As an example, let's predict the future platoon split for the left-handed batter Jim Thome based on his platoon splits during the 2005–7 seasons. During these seasons Thome had a 0.121 platoon split and faced 897 right-handed pitchers. Our predicted future platoon split for Thome is

$$\frac{897}{897+1,000}(.121)+\frac{1,000}{897+1,000}(.022) = .069.$$

This means our best guess is that next season Thome will have an OBP 69 points better against a right-handed pitcher than against a left-handed pitcher.

TABLE 12.1
Predicting Future Platoon Splits

Type of player	Future Estimate of Platoon Split
Right-handed batter	$\dfrac{PAL}{PAL+2200}PS+\dfrac{2200}{PAL+2200}(.013)$
Left-handed batter	$\dfrac{PAR}{PAR+1000}PS+\dfrac{1000}{PAR+1000}(.022)$
Right-handed pitcher	$\dfrac{LHBFP}{LHBFP+700}PS+\dfrac{700}{LHBFP+700}(.023)$
Left-handed pitcher	$\dfrac{RHBFP}{RHBFP+450}PS+\dfrac{450}{RHBFP+450}(.012)$

Note: PS = player's past platoon split; PAL = plate appearances by batter against left-handed pitcher; PAR = plate appearances by batter against right-handed pitcher; RHBFP = right-handed batters faced by pitcher; LHBFP = left-handed batters faced by pitcher.

Source: Tom Tango, Mitchell Lichtman, and Andrew Dolphin, *The Book: Playing the Percentages in Baseball* (Potomac Books, 2007), 377–80.

How Much Can Platooning Help a Team?

TLD show that if a team completely platoons a position in the batting lineup for the entire season (that is, a team always has a left-handed batter hit against a right-handed pitcher and a left-handed batter hit against a right-handed pitcher), they would on average win one more game than they would if they played a single player of comparable overall ability for the whole season. The problem is, of course, that platooning in this matter uses up a roster spot that might be better spent on a pitcher or a good backup fielder.

13

WAS TONY PEREZ A GREAT CLUTCH HITTER?

Tony Perez played first base for the "Big Red Machine" during the 1960s and 1970s and had a lifetime batting average of .279. Such an average does not often lead to a Hall of Fame selection, but in 2000 Perez was elected to the Hall of Fame while some of his contemporaries who have similar statistics (such as Andre Dawson and Dave Parker) have not yet been elected. One reason Perez made it to the Hall of Fame was that his manager, Sparky Anderson, said that Perez was the best clutch hitter he had ever seen. Is there an objective way to determine whether Perez was a great "clutch hitter"?

Let's define a batter to be a great clutch hitter if his performance in important situations tends to be better than his overall season performance. In *Baseball Hacks* Adler defines a clutch situation as one in which a batter comes to the plate during the ninth inning or later and his team trails by one, two, or three runs. Then Adler compared the batter's OBP in these situations to his overall season OBP. If the batter did significantly better during the clutch situations, then we could say the batter exhibited clutch hitting ability. The problem with this approach is that the average batter only encounters ten clutch situations per season, which does not provide enough data to reliably estimate a hitter's clutch ability.[1]

Creating a Benchmark for Expected Clutch Performance

In reality, each plate appearance has a different level of "clutch" importance. When a player goes to the plate to bat when his team is down by a run with two outs in the bottom of the ninth inning, this is obviously a clutch plate appearance, while batting in the top of the ninth when his

---

[1] Adler, *Baseball Hacks*, 345.

team is ahead by seven runs has virtually no "clutch" importance. Recall that the SAGWINDIFF ratings discussed in chapter 8 are based on the relative importance of each plate appearance toward winning or losing a game. Let's define the normalized SAGDIFF rating as the number of points per plate appearance by which a batter exceeds the average rating of 0.

How can we determine whether a player's clutch performance during a season or career was significantly better or worse than his overall ability would indicate? Let's use the player's overall hitting ability (as defined by OBP and SLG) to create a prediction or benchmark for the normalized SAGDIFF rating per plate appearance based on the batter's overall statistics. Then we can say that the player exhibits clutch ability if his actual normalized SAGDIFF rating is significantly higher than what we predicted.

We will restrict our data to players with at least 500 AB during a season. Our dependent variable for each player is his normalized SAGDIFF rating. For example, in 2001 Mo Vaughn's SAGDIFF rating was three points below average per plate appearance. This is our dependent variable. His OBP = .377 and SLG = .498. These are the values of our independent variables. We obtained the following regression result:

$$\text{normalized SAGDIFF rating} = -42.85$$
$$+ 74.12(\text{OBP}) + 38.61(\text{SLG}). \quad (1)$$

As in chapter 3, we see that OBP is roughly twice as important as SLG. This regression explains 81% of the player's variation in SAGDIFF rating. The standard error of this regression is 3.02. This regression shows that most of the normalized SAGDIFF rating is explained by a player's overall hitting ability. The standard error of the regression tells us there is a 95% chance that a player's actual normalized SAGDIFF rating is within two standard errors, or 6.04 points, of our prediction. If a player's normalized SAGDIFF rating is at least 6 points larger than predicted from (1), then he exhibits significant clutch hitting ability; if a player's normalized SAGDIFF rating is more than 6.04 points worse than our prediction from (1), then the player exhibited significantly poor clutch hitting ability.

Tony Perez

Figure 13.1 shows Tony Perez's OBP, SLG, predicted SAGDIFF rating from (1), and actual SAGDIFF rating during the years 1967–75. During all nine years Perez's actual normalized SAGDIFF was larger than his overall hitting

	G	H	I	J	K
6	Year	OBP	SLG	Actual Normalized Sagdif	Predicted Normalized Sagdif
7	1967	328	490	11	0.38
8	1968	338	430	10	−1.19
9	1969	357	526	14	3.92
10	1970	401	589	13	9.62
11	1971	325	438	10	−1.84
12	1972	349	497	13	2.21
13	1973	393	527	13	6.63
14	1974	331	460	8	−0.55
15	1975	350	466	4	1.09

Figure 13.1. Tony Perez was a great clutch hitter.

ability would indicate. In fact, during seven of the nine seasons, Perez's normalized SAGDIFF rating was more than two standard errors better than expected from (1). The facts indeed support Sparky Anderson's assertion that Perez was a great clutch hitter.

During 1967–74 Perez averaged a normalized SAGDIFF rating of 11.5 per season. In contrast, during the prime years of his career (1965–72), Hall of Famer Lou Brock averaged a Normalized SAGDIFF rating of only 5.75. During the eight prime years of his career (1975–82), Hall of Famer George Brett averaged a normalized SAGDIFF rating of 11.8 per season, comparable to Perez's rating. During his prime years of 1984–92, Andre Dawson (who has not yet made the Hall of Fame) averaged a normalized SAGDIFF rating of only 6.5.

During their careers, Dawson and Perez had very similar statistics (Perez's OBP = .341 and SLG = .463; Dawson's OBP = .323 and SLG = .482). The fact that during their peak years Perez averaged 5 more normalized SAGDIFF points per plate appearances than Dawson did is strong evidence that Perez was a much better clutch hitter than was Dawson and gives support to the voters who chose Perez and not Dawson for the Hall of Fame.

The 1969 Mets Revisited

The 1969 Mets exceeded all preseason expectations. In 1968 the Mets won only 73 games, and in 1969 they won 102 games. Part of this surprising success was due to the amazing clutch hitting of Art Shamsky and Ron Swoboda. As shown in figure 13.2, both Shamsky and Swoboda hit much better in the clutch than expected.

	L	M	N	O	P	Q
				Predicted Normalized Sagdif	**Actual Normalized Sagdif**	
6	**Player**	**OBP**	**SLG**			**Year**
7	Shamsky	375	488	3.79	14	1969
8	Swoboda	326	361	−4.75	17	1969

Figure 13.2. Shamsky and Swoboda's 1969 clutch hitting.

Shamsky generated 11.2 more points per plate appearance than expected (over 3.7 standard errors more than expected), while Swoboda generated 21.75 more points per plate appearance than expected (over 7 standard errors more than expected). This shows that during the 1969 season, Shamsky and Swoboda exhibited fantastic clutch hitting.

Does Clutch Hitting Ability Persist from Season to Season?

Over the course of their career, do most players tend to be outstanding or poor clutch hitters? Let's define a batter's clutch performance during a season to be

$$\text{clutch measure} = \text{normalized SAGDIFF} - \text{predicted SAGDIFF from (1)}$$

If a batter exhibits relatively consistent clutch tendencies throughout his career you would expect a positive correlation between the Clutch Measure averaged over a player's even-numbered seasons and his Clutch Measure averaged during his odd-numbered seasons. A significant positive correlation between the even- and odd-year averages would imply that a player who exhibits good (bad) clutch performance during even-numbered seasons would also tend to exhibit good (bad) clutch performance during odd-numbered seasons. The authors of *Baseball between the Numbers* correlated a similar measure of average clutch performance for a batter's even-numbered seasons with a batter's average clutch performance during odd-numbered seasons. They found a correlation of .32.[2] This shows that there is a moderate level of consistency between the clutch performances of a hitter in their even-numbered and odd-numbered season.

[2] Baseball Prospectus Team of Experts, Keri, and Click, *Baseball between the Numbers*, 30.

PITCH COUNT AND PITCHER EFFECTIVENESS

In October 2003 the Red Sox were leading the Yankees 5–2 after seven innings of the seventh and deciding game of the American League Championship Series. Pitcher Pedro Martinez was cruising along and had allowed only two runs. At the start of the eighth inning Martinez got the first batter out, but then Derek Jeter hit a double. Red Sox manager Grady Little went to the mound and talked to Martinez, and then left him in the game. The Yankees promptly tied the game and went on to win in the eleventh inning on a dramatic walk-off home run by Aaron Boone. Grady Little was fired later that week. Most baseball analysts think that one reason Little was fired was that he ignored Martinez's tendency to become, after throwing 100 pitches, a much less effective pitcher. Going into the eighth inning, Martinez was well over the 100-pitch count. Batters facing Martinez after he had thrown fewer than 100 pitches had an OBP of 0.256.[1] This meant that before Martinez had thrown 100 pitches, a batter had a 26% chance of reaching base. After Martinez had thrown more than 100 pitches, batters had a 0.364 OBP, or a 36% chance of reaching base. Since the average OBP is 34%, when Martinez has thrown a lot of pitches he becomes less effective. Grady Little ignored this record, and in all likelihood (since the Yankees had only a 10% chance of winning the game at the start of the eighth inning) his decision cost the Red Sox the 2003 American League championship.

Teams whose front offices' philosophy is data-driven keep records of changes to a pitcher's effectiveness as he throws more pitches. In *The Book*, TLD cleverly analyzed how a pitcher's effectiveness varies with the number of pitches thrown. Using every plate appearance for the 1999–2002 seasons, they analyzed how each hitter performed (after adjusting for the individual abilities of the hitter and pitcher) for every time a pitcher worked through the nine-man batting order.[2] This measure of hitting performance

[1] Adler, *Baseball Hacks*, 358.

was identified as Weighted On-Base Average (WOBA) and was defined by the calculation

$$\frac{.72 \times BB + .75(HBP) + .9(\text{singles}) + .92(\text{reached by error}) + 1.24(2B) + 1.56(3B) + 1.95(HR)}{\text{plate appearances}}.$$

WOBA is a form of Linear Weights (see chapter 3) that is scaled to reflect the average hitter's WOBA of 0.340 (matching the average MLB OBP). Table 14.1 shows the resulting TLD scale.

Therefore, the first time through the order hitters performed eight points below the expected WOBA, while the third time through the order hitters performed eight points better than expected. A pitcher usually finishes going through the order the third time during innings five through seven. The fact that pitchers perform better earlier in the game can be attributed to several factors, including pitcher fatigue and the fact that hitters know what to expect from a pitcher as the game progresses. The fourth time through the batting order hitters performed virtually as expected. This may indicate that any pitcher who makes it to the seventh or eighth inning must have really good stuff, thereby counterbalancing the fatigue effect.

In short, teams would be wise to keep results like those presented in table 14.1 for all hitters each starting pitcher faces, and on an individual basis determine whether a pitcher's performance deteriorates when his pitch count exceeds a certain level.

Pitch Count and Pitcher Injuries

In 1999 Kerry Wood developed a sore elbow, which broke the hearts of Cub fans. During his rookie year in 1998 Wood averaged 112 pitches per game, and during one start he threw 137 pitches. It seems reasonable to assume that starting pitchers who throw a lot of pitches are more likely to develop a sore arm. In *Mind Game*, Steven Goldman describes the link between pitch count and the likelihood a pitcher will develop a sore arm. Keith Woolner and Rany Jazayerli defined Pitcher Abuse Points (PAP for short) for a single start as PAP = max(0,(number of pitches − 100)^3).[3] For example, in any start

---

[2] Tango, Lichtman, and Dolphin, *The Book*, 30.

[3] See Rany Jazayerli, "Pitcher Abuse Points: A New Way to Measure Pitcher Abuse," June 19, 1998, http://www.baseballprospectus.com/article.php?articleid=148.

TABLE 14.1
WOBA vs. Time through Batting Order

	Expected WOBA (Based on Hitters' WOBA)	Actual WOBA	Actual versus Expected WOBA
1	.353	.345	−8 points
2	.353	.354	+1 point
3	.354	.362	+8 points
4	.353	.354	+1 point

Source: Tom Tango, Mitchell Lichtman, and Andrew Dolphin, *The Book: Playing the Percentages in Baseball* (Potomac Books, 2007), 186.

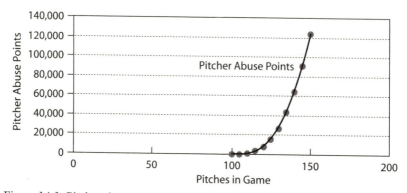

Figure 14.1. Pitcher abuse points as a function of game pitch count. PAP is not a linear function of pitch count.

in which a pitcher throws 100 or fewer pitches, PAP = 0. If a pitcher throws 110 pitches, PAP = 1,000; if a pitcher throws 130 pitches, PAP = 30^3 = 27,000 (see figure 14.1).

Woolner and Jazayerli found that pitchers with a career value of PAP/total pitches exceeding 30 are nearly twice as likely to develop a sore arm as pitchers whose career PAP/total pitches does not exceed 30. For example, a pitcher with five starts throwing 100 pitches in three games and 120 and 130 pitches in the other two games would be a candidate for a sore arm:

$$PAP/\text{total pitches} = (20^3 + 30^3)/550 = 63.6.$$

PAP/total pitches is a tool managers can use to monitor the pitch counts of their starting pitchers to reduce the likelihood of pitcher injury.

15

WOULD TED WILLIAMS HIT .406 TODAY?

In 1941 Ted Williams hit .406. If he were in his prime today (say, the 2006 season), could he still hit around .400? Across the United States arguments similar to the following take place every day: Could Bill Russell dominate Shaq? Who was better: Peyton Manning or Joe Montana? Of course we can't know for sure the answers to these questions. We can, however, use mathematics to determine whether today's players are superior to players from an earlier time.

Let's examine how hitters from the 1940s through the 1980s compare to the hitters in 1941. We will define the level of pitching + defense (PD) in 1941 (PD1941 for short) to be average. If, for example, PD1990 = .10, that would mean a batter hitting against PD1990 would hit 0.10 (or 100 points) higher than a batter hitting against PD1941. If PD1990 = −.10, that would mean a batter hitting against PD1990 would hit 0.10 (or 100 points) lower than a batter hitting against PD1941.

Since PD1941 = 0, simple algebra shows that

$$
\begin{aligned}
\text{PD2005} - \text{PD1941} &= \text{PD2005} \\
&= (\text{PD1942} - \text{PD1941}) + (\text{PD1943} - \text{PD1942}) \\
&\quad + (\text{PD1944} - \text{PD1943}) \\
&\quad + \cdots + (\text{PD2004} - \text{PD2003}) \\
&\quad + (\text{PD2005} - \text{PD2004}).
\end{aligned}
$$

How can we estimate PD1942 − PD1941? Let's assume that the ability of all the 1941 hitters who were still playing in 1942 did not change from 1941 to 1942. Since young players tend to improve with experience and older players tend to lose ability over time, it seems reasonable to assume that the ability of a given cohort of players will not change much from year to year. Given this assumption, suppose the 1941 players who played in 1942 had a batting average (BA) of .260 in 1941 and .258 in 1942. This would indicate that PD1942 was .002 points better than PD1941 (PD1942 − PD1941 = −.002). Thus, in general, PD1942 − PD1941 = BA of the cohort of these hitters in 1942 minus the BA of the cohort of 1941 players in 1941. A similar relationship holds when determining the relative merits of PD during successive years. In the file batting.xlsx, we used some powerful Excel functions (VLOOKUP, IF, OFFSET, SUMIFS, and COUNTIFS) to determine the number of AB and hits in years x and x + 1 for all players who played in both years x and x + 1. For example, we found that players who played in both 1941 and 1942 had the statistics given in table 15.1. Thus PD1942 − PD1941 = −.009, indicating that the 1942 PD was 9 points better than the 1941 PD.

We find the following:

- PD1950 − PD1941 = −.015
- PD1960 − PD1941 = −.022
- PD1970 − PD1941 = −.039
- PD1980 − PD1941 = −.044
- PD1990 − PD1941 = −.057
- PD2005 − PD1941 = −.062

This indicates that 1950 PD is 15 points better than 1941 PD; 1960 PD is 22 points better than 1941 PD; 1970 PD is 39 points better than 1941 PD; 1980 PD is 44 points better than 1941 PD; 1990 PD is 57 points better than PD1941; and 2005 PD is 62 points better than 1941 PD. In 2005 Ted Williams would have faced PD that was 62 BA points better than 1941 PD; he would have hit .406 − .062, or .344 in 2005. Essentially we have found that the strength of MLB pitching and defense has improved by around one BA (62/64) points per year. The greatest improvement occurred in the 1940s (probably because pitching was worse than expected during World War II because many would-be players were serving their

TABLE 15.1
Batting Average for 1941 and 1942 Cohorts of Batters
Who Played Both Years

Year	AB	Hits	BA
1941	71,404	18,822	.264
1942	71,437	18,212	.255

country) and the 1960s (probably because the 1960s saw an influx of great black and Latino pitchers such as Bob Gibson and Juan Marichal).

Our methodology for comparing players of different eras is based on the "Davenport Translations."[1] Similar calculations can be used to predict how minor league prospects will perform in the majors. We will discuss these ideas in chapter 17, and in chapter 50 we will use a more sophisticated mathematical approach to determine whether NBA player quality has improved or declined during the past decade.

[1] See Baseball Prospectus Team of Experts, Keri, and Click, *Baseball between the Numbers*, xxvii.

WAS JOE DIMAGGIO'S 56-GAME HITTING STREAK THE GREATEST SPORTS RECORD OF ALL TIME?

In a beautifully written article, the late paleontologist and lifelong baseball fan Stephen Jay Gould argues that Joe DiMaggio's 56-game consecutive hitting streak is the greatest sports record of all time.[1] In this chapter we will use basic probability and statistics to determine how likely it is that a 56-game hitting streak would ever occur.

In June 1938 Johnny Vander Meer pitched consecutive no-hitters. This has never been done by anyone else. Is this the greatest sports record of all time? After making some reasonable assumptions, basic probability and statistics can help us determine that the occurrence of a 56-game hitting streak was less likely than the occurrence of consecutive no-hitters.

Furthermore, can basic probability and statistics help explain why there have been only seventeen regular season perfect games since 1900? To address these issues we need to study the basic mathematics of rare events.

Calculating the Probabilities of Rare Events: The Poisson Random Variable

Let's consider random variables that always assume a value of 0, 1, 2, and so forth, and the value of the random variable results from the occurrence of rare events. Some examples follow.

- number of accidents a driver has in a year
- number of perfect games during a baseball season
- number of defects in a cell phone

---

[1] Stephen Jay Gould, "The Streak of Streaks," review of *Streak: Joe DiMaggio and the Summer of '41* by Michael Seidel, August 18, 1988, *New York Review of Books*, available at http://www.nybooks.com/articles/4337.

Such random variables are usually governed by the Poisson random variable. Given that a Poisson random variable has a mean λ, then the probability that the random variable equals x is given by $\frac{\lambda^x e^{-\lambda x}}{x!}$. Here x! = (x)(x − 1)x(x − 2) . . . x 1. Thus 3! = 6, 4! = 24, and so forth. The probability that a Poisson random variable with mean λ assumes a value x can be computed in Excel with the formula

$$= POISSON(x, \lambda, False).$$

The Excel function

$$= POISSON(x, \lambda, True)$$

gives the probability that a Poisson random variable with mean λ is less than or equal to x.

For example, if teenage drivers average 0.1 accidents per year, the probability a teen driver has exactly 1 accident during a year is $\frac{.1^1 e^{-.1}}{1!} = .0904.$ This probability can also be calculated in Excel with the formula = POISSON(1, .1, False).

Calculating the Probability of Independent Events

Two events are said to be independent if knowing that one of the events occurred tells you nothing about the likelihood the other event will occur. For example, the Chicago Cubs winning the World Series and the Chicago Bears winning the Super Bowl during the same season are independent events because if you knew that the Cubs won the World Series (we can dream!) you would not change your view about the probability that the Bears would win the Super Bowl. To calculate the probability that multiple independent events will occur at the same time, we simply multiply the probability of the individual events.

For example, if the average major leaguer has an OBP of 0.34, what is the probability that an average pitcher would pitch a perfect game? A pitcher must retire 27 consecutive batters to be credited with pitching a perfect game. Assuming that the events that batter I reaches base are independent and all have a probability of .34, then the probability of a pitcher throwing a perfect game would be $(1 − .34)^{27} = .0000134$, or around one chance in 74,500.

What Is the Probability for the Seventeen
Perfect Games Pitched since 1900?

There have been seventeen regular season perfect games since 1900. Since the probability that a starting pitcher will throw a perfect game is .0000134, what is the likelihood that seventeen regular season perfect games would have occurred since 1900?

There have been nearly 173,000 regular season games since 1900. Each game presents two opportunities for a perfect game. Since each pitcher's start will yield either 0 or 1 perfect game, *the probability that a pitcher will throw a perfect game is equal to the expected number of perfect games a pitcher throws in a game*. For any set of random variables, the expected value of the sum of the random variables is equal to the sum of the expected values of the individual random variables. This leads us to the conclusion that the expected number of perfect games since 1900 should be $2 \times (173,000) \times (.0000134) = 4.64$. Because perfect games since 1900 are rare, they should follow a Poisson random variable. Then the probability of the occurrence of at least seventeen perfect games would be given by $1 - \text{POISSON}(16, 4.64, \text{True}) = .000008$, or eight chances in one million. Clearly our model predicts significantly fewer perfect games than occurred. What might be wrong with our assumptions?

We assumed every batter had an OBP of exactly 0.34. Batters like Barry Bonds have an OBP much higher than 0.34, while a batter who hit against Sandy Koufax had an OBP much less than 0.34. Let's assume that in half of all games each hitter has an OBP of 0.29 and in half of all games each hitter has an OBP of 0.39. This averages out to our observed OBP of 0.34. Then the probability that a starting pitcher will throw a perfect game would (by the Law of Conditional Expectation introduced in chapter 6) be equal to

$$\text{(probability starting pitcher yields OBP of .29)}(1 - .29)^{27}$$
$$+ \text{(probability starting pitcher yields an OBP of .39)}$$
$$\times (1 - .39)^{27} = .5 \times (1 - .29)^{27} + .5(1 - .39)^{.27}$$
$$= .0000489883.$$

Then our expected number of perfect games since 1900 becomes $346,000(0000489883) = 16.95$, which is clearly consistent with the actual number of 17 perfect games.

Our model might also be in error because we treated each plate appearance as a separate event, equally capable of resulting in the player making it on base. If a pitcher is doing well and retires the first, say, ten batters, we might lower our estimate of how many later hitters reach base, thus violating our independence assumption.

How Unusual Was the 56-Game Hitting Streak?

To determine the probability of a 56-game hitting streak occurring during 1900–2006, let's make the following assumptions.

- Only batters with at least 500 AB in a season can have a 56-game hitting streak.
- We will not include hitting streaks that carry over between seasons (in theory these count as hitting streaks).
- Batters with over 500 AB during a season averaged 3.5 AB per game, so we will assume each batter had 3 AB in half his games and 4 AB in the other half.

We can now estimate for each batter since 1900 with at least 500 AB during a season the chance he could have had a 56-game hitting streak. We will soon see that overall, there is roughly a 2% chance that at least one 56-game hitting streak could have occurred.

To illustrate, let's consider a batter who hit .316 in 1900 (a 154-game season). We can compute the following:

$$\text{probability batter gets a hit in 3 AB game}$$
$$= 1 - \text{probability he fails to get a hit on all 3 AB}$$
$$= 1 - (1 - .316)^3 = .678.$$

Similarly,

$$\text{probability batter gets a hit in 4 AB game}$$
$$= 1 - (1 - .316)^4 = .781.$$

Let's assume that each player will have 28 three AB games and 28 four AB games during a 56-game sequence (this assumption has little effect on our final result). Then the probability that a batter will get a hit during each of 56 consecutive games is

$$(\text{probability a batter gets a hit in 3 AB game})^{28}$$
$$\times (\text{probability batter gets a hit in 4 AB game})^{28}$$
$$= .00000002 \text{ (roughly 2 in 100 million)}.$$

How many chances does a batter get to start a 56-game streak? He can start the streak during the first game of the season, as well as during any of the first 99 games of the season for which he was hitless during the previous game. We approximate the probability that a batter is hitless in a game by using the average of the probability he is hitless in a 3 AB game and the probability he is hitless in a 4 AB game. Therefore,

$$\text{probability of hitless game} = .5 \times (1 - .68)$$
$$+ .5\,(1 - .781) = .27.$$

Thus, the batter has, on average, $1 + 98 \times (.27) = 27.46$ opportunities to start his 56-game hitting streak. This implies that the expected number of 56-game hitting streaks by a batter during the season is 27.46 $(.00000002) = .0000055$.

In file Dimaggio.xlsx the expected number of 56-game hitting streaks over all batters with at least 500 AB during the 1900–2006 seasons are summed up, generating an expected total of .024 56-game hitting streaks. Using the Poisson random variable, the probability of at least one 56-game hitting streak occurring is

$$1 - \text{POISSON}(.0, .024, \text{True}) = .024.$$

Our calculations show that given all the opportunities for a 56-game hitting streak to occur, such a streak is highly unlikely, but certainly not impossible. Another way to put the likelihood of a 56-game hitting streak in perspective is to determine how many years a batter would have to play before he had a 50% chance of having a 56-game hitting streak. Figure 16.1 shows the results of such calculations, assuming the batter had 4 AB per game and a batting average of .300 to .500. For example, we find that a .400 hitter would have to play for 120 seasons to have a 50% chance of having a 56-game hitting streak.

How Unusual Is It to Pitch Consecutive No-Hitters?

What is the probability that at least one starting pitcher (since 1900) would have pitched consecutive no-hitters? To answer this question we will make the following assumptions:

- All games are started by pitchers who start exactly 35 games during a season.
- Since 1900, 0.00062 of all pitchers starting a game have pitched a no-hitter. We therefore assume the probability that each game started will result in a no-hitter is .00062.

	I	J
10	**Batting Average**	**Seasons**
11	0.3	134514.9521
12	0.31	53294.89576
13	0.32	22381.65641
14	0.33	9926.429104
15	0.34	4633.715654
16	0.35	2269.681638
17	0.36	1163.246945
18	0.37	622.1841078
19	0.38	346.4659034
20	0.39	200.4140313
21	0.4	120.1772114
22	0.41	74.55979981
23	0.42	47.77425381
24	0.43	31.56151664
25	0.44	21.46397597
26	0.45	15.00394473
27	0.46	10.76553103
28	0.47	7.91818304
29	0.48	5.962542423
30	0.49	4.59130992
31	0.5	3.611179813

Figure 16.1. Seasons needed to have a 50% chance of getting a 56-game hitting streak.

To determine the expected number of times a pitcher will throw consecutive perfect games, we first determine the probability that a pitcher making 35 consecutive starts during a season throws consecutive no-hitters. This probability is given by

(probability the first of the two no-hitters is game 1)
+ (probability the first of the two no-hitters is game 2)
+ . . . (probability the first of the two no-hitters is game 34).

Now the probability the pitcher throws consecutive no-hitters and the first of the two no-hitters is game $1 = (.00062)^2 = .00000038$. (See file Nohitters.xls.) The probability the pitcher throws consecutive no-hitters and the first no-hitter is game $i = (1 - .00062)^{i-1}(.00062)^2$, which is approximately .000000038. Therefore, the probability that a pitcher with 35 starts will pitch consecutive no-hitters during a season is approximately $34 \times (.000000038) = .0000128$. Our computations assume that if a pitcher throws two consecutive no-hitters, he has thrown no prior no-hitters during the season. We performed a Monte Carlo simulation to confirm that this assumption has little effect on the end result of our probability calculations.

In 1900, 1,232 games were played and pitchers made $1,232 \times 2 =$ 2,464 starts. At 35 starts per pitcher, this would imply there were around 2,464/35 (around 70) starting pitchers that year. Thus we find the expected number of times consecutive no-hitters would be pitched during 1900 to be $70 \times (.0000128) = .00089$.

Adding together the expected number of consecutive no-hitter occurrences for the years 1900–2007, we find the expected number of consecutive no-hitters during 1900–2007 to be 0.126. Therefore, our model implies that the probability that at least one starting pitcher would throw consecutive no-hitters is $1 -$ POISSON(0,.126,True) = 11.8%.

Another way to demonstrate how rare consecutive no-hitters are is to determine how many years (with 35 starts per year) the greatest no-hit pitcher of all time (Nolan Ryan) would have to pitch before he had a 50% chance of throwing consecutive no-hitters. Ryan pitched 6 no-hitters in 773 starts, so the chance that any of his starts would result in a no-hitter is 6/773 = .00776. This yields a .0018 chance of consecutive no-hitters during a season. Ryan would have needed to pitch 384 seasons to have a 50% chance of throwing consecutive no-hitters.

In summary, we find that consecutive no-hitters and a 56-game hitting streak are both highly unlikely events but not beyond the realm of possibility.

A Brainteaser

We close this chapter with a brainteaser. A starting pitcher pitches a complete game and the game is nine innings. What is the minimum number of pitches the starting pitcher could have thrown? The answer is 25. The starting pitcher is pitching for the visiting team and allows only one runner to reach base (on a home run) and loses 1–0. The starting pitcher does not need to pitch the ninth inning, so he throws $8 \times 3 + 1 = 25$ pitches.

17

MAJOR LEAGUE EQUIVALENTS

Major league general managers must decide every year whether a promising minor league player is ready to be brought up to the major league team. Of course, the minor league player faces inferior pitching in the minors, so he is not expected to duplicate his minor league statistics when he is brought up to the majors. In 1985 Bill James developed Major League Equivalents to help major league front office personnel determine whether a minor leaguer is ready for the majors.

The Excel file mle.xls gives the OBP for a set of hitters whose last minor league year was played at the AAA level. These hitters played in either the American Association (AA), International League (INT), or Pacific Coast League (PCL). The file also gives their OBP during their first major league season.

Suppose we know a batter, Joe Hardy, had an OBP of 0.360 in AAA. If we bring Joe up to the major leagues, what OBP can we expect? Using the available data for the INT, we learn that batters had an OBP that averaged 90% of their last minor league OBP during their first year in the majors after their last year (or part of year) in the INT. Batters who had played in the PCL averaged 88% of their last minor league OBP during their first year in the majors. Thus, the major league equivalent of an AAA minor league OBP would be roughly 0.89 times the minor league OBP. We would therefore predict that Joe would achieve a "major league equivalent" OBP of .89 × (.360) = .320 in the major leagues.

Expert sabermetricians know that major league equivalents should be adjusted for the minor league park, the major league park, and the quality of pitching faced in the minor league. For example, Tucson and Albuquerque are known to be hitters' parks, so batters who play for these teams would have their major league equivalent OBPs reduced.[1] Similarly, if a

---

[1] See http://www.baseballamerica.com/today/features/040408parkfactors.html.

batter were being called up to a team like the Dodgers, who play in a park in which it is more difficult than average to reach base, their projected major league equivalent should be reduced as well.

I did not collect data on slugging percentages, but let's suppose slugging percentage also drops off around 11% when a player goes from AAA to the major leagues. Recall from chapter 2 that Bill James's original Runs Created Formula is (for all intents and purposes) computed by multiplying OBP and SLG. Thus we would expect an AAA minor leaguer to retain $.89^2 = .78$, or 78% of their minor league Runs Creating ability and lose around 22%.

PART II

FOOTBALL

18

WHAT MAKES NFL TEAMS WIN?

NFL teams want to win games. Is it more important to have a good rushing attack or a good passing attack? Is rushing defense more important than passing defense? Is it true that turnovers kill you? During the early 1960s statistician Bud Goode studied what makes a team win. He found that passing yards per attempt (PY/A) on both offense and defense were the most important factors in predicting an NFL team's success. This is intuitively satisfying because PY/A is more of a measure of efficiency than total yards passing. Since we divide by pass attempts, PY/A recognizes that passing plays use up a scarce resource (a down).

Using team statistics from the 2003–6 season we will run a regression to try to predict each team's scoring margin (points for − points against). The independent variables used are listed below. All our independent variables will prove to have p-values less than .05. (See file NFLregression.xls.)

- Team offense PY/A. We include sacks in pass attempt totals, and yards lost on sacks are subtracted from yards passing.
- Team defense-passing yards per attempt: (DPY/A). We include sacks in pass attempt totals, and yards lost on sacks are subtracted from yards passing.
- Team offense rushing yards per attempt (RY/A).
- Team defense rushing yards per attempt (DRY/A).
- Turnovers committed on offense (TO).
- Defensive turnovers (DTO).
- Differential between penalties committed by team and penalties committed by its opponents (PENDIF).
- Return TDs − return TDs by opponent (RET TD) (includes TDs scored on fumbles, interceptions, kickoffs, and punts).

Again our dependent variable for each team is total regular season Points For − Points Against.

After running our regression (and eliminating the intercept, which had a non-significant p-value) we obtain the output shown in figure 18.1.

The RSQ of .87 tells us that the following equation explains 87% of the variation in team scoring margin. The standard error of 35 means that our equation predicts the scoring margin for 95% of the teams within 70 points (or 4.25 points per game).

$$\text{predicted team scoring margin for season}$$
$$= 3.17(\text{RET TD}) - .06(\text{PENDIF}) + 61.67(\text{PY/A})$$
$$+ 26.44(\text{RY/A}) - 2.77(\text{TO}) - 67.5(\text{DPY/A})$$
$$- 22.79(\text{DRY/A}) + 3.49(\text{DTO}).$$

From this regression we learn that (after adjusting for the other independent variables):

- an extra TD on return is worth 3.17 points;
- an extra PY/A is worth 61.67 points;
- an extra RY/A is worth 26.44 points;
- an extra TO costs 2.77 points;
- an extra PY/A costs 67.5 points;
- an extra RY/A costs 22.79 points;
- an extra forced TO produces 3.49 points; and
- an extra yard in penalties costs 0.06 points.

The coefficients for offensive and defensive passing efficiency are almost triple the coefficients for offensive and defensive rushing efficiency. This is consistent with Goode's finding that passing efficiency is the key driver of success in the NFL. The standard deviation of team PY/A is 0.83 yards and the standard deviation of RY/A is 0.55 yards. This means that if an average passing team were to move to the 84th percentile in PY/A (one standard deviation above average) their performance would improve by $0.83 \times 61.67 = 52$ points; if an average rushing team were to move up to the 84th percentile in RY/A, their performance would improve by $0.5(26.44) = 13.22$ points. If it costs the same (in terms of percentile ranking) to improve passing and rushing offense, a team is better off trying to improve its passing offense. Our results give little credence to the belief of so-called experts that you need a good ground game to set up your passing game. In fact the correlation between PY/A and RY/A is only 0.10.

	A	B	C	D	E	F
1	**SUMMARY OUTPUT**					
2						
3	*Regression Statistics*					
4	**Multiple R**	0.934530278				
5	**R Square**	0.873346841				
6	**Adjusted R Square**	0.857625407				
7	**Standard Error**	35.18940561				
8	**Observations**	128				
9						
10	**ANOVA**					
11		*df*	*SS*	*MS*	*F*	*Significance F*
12	**Regression**	8	1024650.688	128081.336	103.4336825	6.80921E–50
13	**Residual**	120	148595.312	1238.294267		
14	**Total**	128	1173246			
15						
16		*Coefficients*	*Standard Error*	*t Stat*	*P-value*	
17	**Inter-ceptions**	0	#N/A	#N/A	#N/A	
18	**Returned for TD**	3.172538521	1.2791899	2.480115362	0.014520952	
19	**OFF pen./ DEF pen.**	–0.056152517	0.019530575	–2.875108178	0.004779821	
20	**Neg. Yds. per Att.**	61.67756012	3.957309602	15.58573029	1.28665E–30	
21	**Yards per Att.**	26.44027931	6.303341114	4.194645162	5.26202E–05	
22	**OFF Turnovers**	–2.771793664	0.473496142	–5.853888596	4.27337E–08	
23	**DEF Yds. per Pass Att.**	–67.49861999	5.396830578	–12.50708522	1.73149E–23	
24	**DEF Yds. per Rush Att.**	–22.78529422	7.309963853	–3.117018727	0.002286651	
25	**DEF Turnovers**	3.48978828	0.50392494	6.925214462	2.3113E–10	

Figure 18.1. Results of NFL regression. Note that all independent variables have p-values less than 0.02 and are significant.

	P	Q	R	S	T	U	V	W	X
8		Returned for TD	Penalty Differential	OFF Pass Yds. per Att.	OFF Rush Yds. per Att.	OFF Turnovers	DEF Pass Yds. per Att.	DEF Rush Yds. per Att.	DEF Turnovers
9	Returned for TD	1	−0.10949	0.07646	0.15469	−0.31181	−0.21578	−0.08204	0.37086
10	Penalty Differential	−0.10949	1	−0.17464	−0.03113	0.13023	0.06286	−0.1739	−0.10505
11	OFF Pass Yds. per Att.	0.07646	−0.17464	1	0.0999	−0.44827	0.01083	0.17965	0.15363
12	OFF Rush Yds. per Att.	0.15469	−0.03113	0.0999	1	−0.31013	0.06501	0.09502	−0.03066
13	OFF Turnovers	−0.31181	0.13023	−0.44827	−0.31013	1	−0.02521	−0.09825	−0.13314
14	DEF Pass Yds. per Att.	−0.21578	0.06286	0.01083	0.06501	−0.02521	1	0.2759	−0.29984
15	DEF Rush Yds. per Att.	−0.08204	−0.1739	0.17965	0.09502	−0.09825	0.2759	1	−0.14402
16	DEF Turnovers	0.37086	−0.10505	0.15363	−0.03066	−0.13314	−0.29984	−0.14402	1

Figure 18.2. Correlation matrix for NFL independent variables.

Another way to show the importance of the passing game is to run a regression to predict scoring margin using only PY/A and DPY/A. Together these variables explain 70% of the variation in team scoring margin. In contrast, predicting team scoring margin from RY/A and DRY/A explains only 6% of the variation in team scoring margin.

The offensive and defensive turnover coefficients average out to 3.13 points. This would seem to indicate that a turnover is worth 3.13 points. Note, however, that we accounted for the effect of return TDs in RET TD. If we drop this independent variable from our model we find that a defensive turnover is worth 3.7 points and an offensive turnover costs 3.4 points. This indicates that a turnover is worth approximately the average of 3.7 and 3.4, or 3.55 points.

The key to continued success in the NFL is understanding how money spent on players changes the values of the independent variables. For example, if an all-pro receiver costs $10 million and an all-pro linebacker costs $10 million, which expenditure will improve team performance more? This is difficult to answer. The linebacker will have a big impact on DTO, DRY/A, and DPY/A, while the receiver will probably only effect PY/A and TO. The front office personnel who can determine which free agent has more impact will be a key contributor to his team's success.

Does a Good Rushing Attack Set Up
the Passing Game?

Most fans (myself included) believe that a good rushing attack helps set up
the passing attack. If this is the case, we would expect to see a strong positive correlation between PY/A and RY/A. Figure 18.2 shows the correlation between our independent variables. Note that PY/A and RY/A have
only a .10 correlation, so a good ground does not seem to lead to a good
passing game. Perhaps this is because teams that pay high salaries to their
quarterback and wide receivers have little money left to pay for running
backs.

19

WHO'S BETTER, TOM BRADY
OR PEYTON MANNING?

Most American men spend a good deal of time arguing about who are the best quarterbacks in the NFL. For example, is Tom Brady better than Peyton Manning? The NFL quarterback rating system works as follows.

> First one takes a quarterback's completion percentage, then subtracts 0.3 from this number and divides by 0.2. You then take yards per attempts subtract 3 and divide by 4. After that, you divide touchdowns per attempt by 0.05. For interceptions per attempt, you start with 0.095, subtract from this number interceptions per attempt, and then divide this result by 0.04. To get the quarterback rating, you add the values created from your first four steps, multiply this sum by 100, and divide the result by 6. The sum from each of your first four steps cannot exceed 2.375 or be less than zero.[1]

This formula makes quantum mechanics or Fermat's Last Theorem seem simple. (The NCAA also has its own incomprehensible system for ranking quarterbacks.)

To summarize, a quarterback's rating is based on four statistics:

- completion percentage (completions per passing attempts)
- yards gained per pass attempt (yards gained by passes) per (passing attempts)
- interception percentage (interceptions per passing attempts)
- TD pass percentage (TD passes per passing attempts).[2]

Like the antiquated *Fielding* percentage metric in baseball, this formula makes no sense. Note that all four statistics are given equal weight in the ranking formula. There is no reason why, for example, completion percent-

---

[1] Berri, Schmidt, and Brook, *Wages of Win*, 167.
[2] Ibid., 167.

age and yards per pass attempt should be given equal weight. In fact, because an incomplete pass gains 0 yards, completion percentage is partially accounted for by yards per pass attempt. In the TD per pass attempt percentage portion of the formula, a TD pass of 1 yard is given as much weight as a touchdown pass of 99 yards. (The 99-yard pass pumps up yards per pass attempt, but it is obvious how arbitrary the system is.) Of course, any system for rating quarterbacks based on individual passing statistics is actually reflecting the entire team's passing game, which is influenced by the quality of the team's receivers and its offensive line. With this caveat, however, we can try to develop a simpler rating system for quarterbacks.

In *Wages of Wins*, Berri, Schmidt, and Brook rate quarterbacks' passing performance using the following formula.

$$\text{quarterback rating} = \text{all yards gained} - 3(\text{passing attempts}) - 30(\text{interceptions}).[3] \tag{1}$$

The coefficients in this rating system are derived from a regression that attempts to predict a team's scoring margin based on its yards gained, number of offensive plays, and turnovers. Note the -3(passing attempts) term penalizes a quarterback for using a scarce resource—a down.[4]

We will base our quarterback's score on Brian Burke's work.[5] In 2007 Burke ran a regression (similar to the one shown in chapter 17) to predict team games won during the 2002–6 NFL seasons. Burke used the following independent variables:

TRUOPASS = offense (passing yards allowed − sack yards)/passing plays

TRUDPASS = defense (passing yards allowed − sack yards)/passing plays

ORUNAVG = offense rushing yards allowed/rushing attempts

DRUNAVG = defense rushing yards allowed/rushing attempts

OINTRATE = interceptions thrown/passing attempts

DINTRATE = interceptions made/passing attempts

OFUMRATE = offensive fumbles/all offensive plays (not fumbles lost)[6]

DFFRATE = defensive forced fumbles/all defensive plays

---

[3] Ibid., 173.

[4] A team is allowed only four downs to gain ten yards. Therefore, downs are scarce resources and the use of a down should be penalized.

[5] See his great Web site, http://www.bbnflstats.com/.

[6] "Not fumbles lost" means all fumbles are used, not just the fumbles on which the team loses possession of the ball.

	D	E	F
8	**Variable**	**Coefficient**	**p-value**
9	const	5.2602	0.06281
10	TRUOPASS	1.54337	<0.00001
11	TRUDPASS	−1.66731	<0.00001
12	ORUNAVG	0.91979	0.00071
13	DRUNAVG	−0.553532	0.04842
14	OINTRATE	−50.0957	0.0012
15	DINTRATE	83.6627	<0.00001
16	OFUMRATE	−63.9657	0.00053
17	DFFRATE	78.6917	0.0001
18	PENRATE	−4.48514	0.01283
19	OPPPENRATE	6.5826	0.0004

Figure 19.1. Regression results for predicting team wins. The quarterback influences TRUOPASS (analogous to PY/A in chapter 17) and OINTRATE.

PENRATE= each team's penalty yards/all plays, offensive and defensive

OPPPENRATE= opponent's penalty yards/all plays, offensive and defensive

The results of his regression are shown in figure 19.1.

We will simply rate each 2007 NFL quarterback using the coefficients from Burke's regression. This should rate quarterbacks according to how their team's passing game favorably impacted the number of games their team won. Therefore our quarterback rating is 1.543(TRUOPASS) − 50.0957(OINTRATE).

Figure 19.2 shows the NFL's quarterback ratings, our ratings, and the ratings from *Wages of Wins* for the 2007 season. We normalized our quarterback scores to average out to 100 by multiplying each QB's rating. Our results are shown in figure 19.2.

All three methods of rating quarterbacks rank Tom Brady first by a wide margin. Both our system and *Wages of Wins* rank Ben Roethlisberger much lower than his NFL rating of second place. This is in part because we count sacks and Roethlisberger was sacked often.

The three rating systems are remarkably consistent. The correlation between our ratings and the NFL ratings is .91. The correlation between the NFL ratings and the *Wages of Wins* ratings is .84. Finally, the correlation

	D	E	F	G	H	I	J	K	L	M
6	Player	Comp.	Attempts	%	Yards	Yards/ Pass Att.	Longest Pass	TD Passes	TD Pass %	Inter- ceptions
7	T. Brady	398	578	68.9	4806	8.32	69	50	8.7	8
8	B. Roethlisberger	264	404	65.3	3154	7.81	83	32	7.9	11
9	D. Garrard	208	325	64	2509	7.72	59	18	5.5	3
10	P. Manning	337	515	65.4	4040	7.85	73	31	6	14
11	T. Romo	335	520	64.4	4211	8.1	59	36	6.9	19
12	B. Favre	356	535	66.5	4155	7.77	82	28	5.2	15
13	J. Garcia	209	327	63.9	2440	7.46	69	13	4	4
14	M. Hasselbeck	352	562	62.6	3966	7.06	65	28	5	12
15	D. McNabb	291	473	61.5	3324	7.03	75	19	4	7
16	K. Warner	281	451	62.3	3417	7.58	62	27	6	17
17	D. Brees	440	652	67.5	4423	6.78	58	28	4.3	18
18	J. Cutler	297	467	63.6	3497	7.49	68	20	4.3	14
19	M. Schaub	192	289	66.4	2241	7.75	77	9	3.1	9
20	C. Palmer	373	575	64.9	4131	7.18	70	26	4.5	20
21	C. Pennington	179	260	68.8	1765	6.79	57	10	3.8	9
22	S. Rosenfels	154	240	64.2	1684	7.02	53	15	6.3	12
23	D. Anderson	298	527	56.5	3787	7.19	78	29	5.5	19
24	P. Rivers	277	460	60.2	3152	6.85	49	21	4.6	15
25	J. Kitna	355	561	63.3	4068	7.25	91	18	3.2	20
26	J. Campbell	250	417	60	2700	6.48	54	12	2.9	11
27	J. Harrington	215	348	61.8	2215	6.37	69	7	2	8
28	D. Huard	206	332	62	2257	6.8	58	11	3.3	13
29	B. Griese	161	262	61.5	1803	6.88	81	10	3.8	12
30	K. Boller	168	275	61.1	1743	6.34	53	9	3.3	10
31	E. Manning	297	529	56.1	3336	6.31	60	23	4.3	20
32	V. Young	238	382	62.3	2546	6.67	73	9	2.4	17
33	C. Lemon	173	309	56	1773	5.74	64	6	1.9	6
34	T. Jackson	171	294	58.2	1911	6.5	71	9	3.1	12
35	T. Edwards	151	269	56.1	1630	6.06	70	7	2.6	8
36	M. Bulger	221	378	58.5	2392	6.33	40	11	2.9	15
37	B. Croyle	127	224	56.7	1227	5.48	35	6	2.7	6
38	R. Grossman	122	225	54.2	1411	6.27	59	4	1.8	7
39	K. Clemens	130	250	52	1529	6.12	56	5	2	10

Figure 19.2. NFL quarterback ratings, regression ratings, and *Wages of Wins* ratings. See file Qb2007statts.xls. Source: David Berri, Martin Schmidt, and Stacey Brook, *The Wages of Wins: Taking Measure of the Many Myths in Modern Sport* (Stanford University Press, 2006), and equation (1) on page 133.

	N	O	P	Q	R	S	T	U	V
6	Inter-ception %	# of Sacks	Yards/ Sacks	NFL rating	Net Yards/ Attempt	Regression Rating	Our rank	"Wages of Wins" Rating	"Wages of Wins" Rank
7	1.4	21	128	117.2	7.809682805	157.4433428	1	2769	1
8	2.7	47	347	104.1	6.223946785	114.4701377	15	1471	13
9	0.9	21	99	102.2	6.965317919	142.8426456	2	1381	14
10	2.7	21	124	98	7.305970149	137.6311052	4	2012	3
11	3.7	24	176	97.4	7.417279412	133.0674385	6	2009	4
12	2.8	15	93	95.7	7.385454545	138.6378608	3	2055	2
13	1.2	19	104	94.6	6.751445087	136.1807663	5	1282	15
14	2.1	33	204	91.4	6.322689076	120.7515008	11	1821	6
15	1.5	44	227	89.9	5.99032882	117.8050112	14	1563	11
16	3.8	20	140	89.8	6.957537155	122.5319193	9	1494	12
17	2.8	16	109	89.4	6.458083832	118.7872701	13	1879	5
18	3	27	153	88.1	6.769230769	124.0581907	8	1595	9
19	3.1	16	126	87.2	6.93442623	126.899612	7	1056	18
20	3.5	17	119	86.7	6.777027027	120.7519375	10	1755	7
21	3.5	26	178	86.1	5.548951049	94.46467906	24	637	24
22	5	6	48	84.8	6.650406504	107.6221931	18	586	28
23	3.6	14	109	82.5	6.798521257	120.5173998	12	1594	10
24	3.3	22	163	82.4	6.201244813	109.8164358	16	1256	16
25	3.6	51	320	80.9	6.124183007	106.0830291	19	1632	8
26	2.6	21	110	77.6	5.913242009	108.5140549	17	1056	18
27	2.3	32	192	77.2	5.323684211	97.9783104	21	835	20
28	3.9	36	234	76.8	5.497282609	90.58019661	27	763	22
29	4.6	15	114	75.6	6.097472924	98.56502548	20	612	25
30	3.6	24	159	75.2	5.297658863	88.39108325	29	546	30
31	3.8	27	217	73.9	5.60971223	93.6814062	26	1068	17
32	4.5	25	157	71.1	5.86977887	94.38580688	25	815	21
33	1.9	25	166	71	4.811377246	89.79076516	28	591	27
34	4.1	19	70	70.8	5.881789137	97.42139672	22	612	25
35	3	12	105	70.4	5.427046263	95.32841213	23	547	29
36	4	37	269	70.3	5.115662651	81.71690384	31	697	23
37	2.7	17	101	69.9	4.67219917	81.25461095	32	324	33
38	3.1	25	198	66.4	4.852	82.32478329	30	451	31
39	4	27	138	60.9	5.02166065	79.70476865	33	398	32

Figure 19.2. (*cont.*)

between our ratings and the *Wages of Win* ratings is .87. It appears that a simpler formula that ties a quarterback's rating to the extent to which his team's passing game creates wins is more useful than the NFL's arbitrary, unnecessarily complex formula.

We will see in chapter 22 that a much better way of evaluating a team's passing attack is available, by calculating how each play changes the expected number of points by which the team wins the game. Unfortunately, however, what is really needed is a way to decompose the effectiveness of a team's passing attack and understand what fraction of a team's passing effectiveness can be attributed to the quarterback, the receivers, and the offensive line. Since the starting offensive line and the starting quarterback are usually in for almost every play, this is very difficult. In chapter 30 we will show that the fact that almost every NBA player sits out at least 17% of each game makes it fairly simple to partition a team's success (or lack thereof) among its players.

20

FOOTBALL STATES AND VALUES

In chapter 8 we discussed how the inning, score margin, outs, and runners on bases were sufficient data when trying to determine whether a baseball team would win a game (assuming two equal teams were playing.) For example, if a team is down by three runs in the top of the seventh inning with two outs and the bases loaded, it has a 15% chance of winning the game. We call the inning, score margin, outs, and runners on bases the state of the baseball game. Once we know the state of the game and have evaluated the chance of winning in each state, we can analyze strategies such as bunting or evaluate (as shown in chapter 8) batters and pitchers based on how they change the team's chance of winning the game.

Football States

If we can define a state for football that is sufficient to determine the chance of winning a game (assuming two equal teams are playing), then we can analyze how plays affect a team's chance of winning. We can then use this information to rate running backs, quarterbacks, and wide receivers. For example, we might find that when running back A carries the ball, on average he adds 0.1 points per carry, while running back B adds 0.3 points per carry. This would indicate that (assuming offensive lines of equal quality) that running back B is better. Comparing running backs based on points added per carry would be a better measure than comparing running backs based on the current metric of yards per carry. We can also use football states to evaluate strategic decisions such as when to go for a two-point conversion after a touchdown (see chapter 24), when to punt on fourth down or go for it, and when to try a field goal on fourth down or go for it, as well as the run-pass mix on first down (see chapter 21).

The state of a football game at any time is specified by the following quantities:

- yard line
- down
- yards to go for first down
- score differential
- time remaining in game

For example, if there are ten minutes to go in the second quarter and a team has third and 3 on its 28-yard line and is down by seven points, the team would like to know its probability of victory.

In baseball the number of states is manageable (several thousand), but in football there are millions of possible states. To simplify analysis, most analysts assume that a team's goal is to maximize, from the current time onward, the expected number of points by which it beats its opponent. We assume the game is of infinite length, meaning a very long game (e.g., 1,000 minutes long). This eliminates the need to know the time remaining in the game.[1] Of course, near the end of the game if a team is down by two points its goal is not to maximize expected points scored but simply to maximize its chance of kicking a field goal. Therefore our assumption of an infinite-length game will not be valid near the end of the game or the end of the first half, but for most of the game "an expected points margin maximizer" will choose decisions that maximize a team's chance of victory. To simplify the state, we will assume that the state in a football game is specified by down, yards to go for first down, and yard line. This simplification still leaves us (assuming we truncate yards to go for a first down at 30) with nearly $4 \times 99 \times 30 = 11,880$ possible states.

Football analysts define the value of a state as the margin by which a team is expected to win (from that point onward) in a game of infinite duration given the team has the ball on a given yard line, as well as the down and yards to go situation. These values are difficult to estimate. Carter and Machol, Romer, and Footballoutsiders.com have all estimated these values for first-and-10 situations.[2] Cabot, Sagarin, and Winston (CSW) have estimated state values for each yard line and down and yards to go situations.[3] A sampling of the values for several first-and-10 situations is given in table 20.1.

[1] The Web site Footballcommentary.com and the ZEUS computer system incorporate time remaining into decisions such as whether to go for it on fourth down or go for a one- or two-point conversion.

[2] Carter and Machol, "Operations Research on Football"; Romer, "It's Fourth Down and What Does the Bellman Equation Say?"

[3] Vic Cabot, Jeff Sagarin, and Wayne Winston, "A Stochastic Game Model of Football," unpublished manuscript, 1981.

TABLE 20.1
NFL State Values

Yard Line	Carter and Machol	Cabot, Sagarin, and Winston	Romer*	Football Outsiders.com*
5	−1.25	−1.33	−0.8	−1.2
15	−0.64	−0.58	0	−0.6
25	0.24	0.13	0.6	0.1
35	0.92	0.84	1.15	0.9
45	1.54	1.53	1.90	1.2
55	2.39	2.24	2.20	1.9
65	3.17	3.02	2.8	2.2
75	3.68	3.88	3.30	3.0
85	4.57	4.84	4.0	3.8
95	6.04	5.84	4.90	4.6

Note: The 5-yard line is 5 yards from the goal line while the 95-yard line is 5 yards from the opponent's goal line.
 * Read off graph, so approximate.

Carter and Machol used data from the 1969 NFL season to estimate the value of first and 10 on only the 5-, 15-, 25-, . . . 95-yard lines. They assume the value to a team receiving a kickoff after a scoring play was 0. CSW analyzed the value for each down, yards to go (less than or equal to 30), and yard-line situation by inputting the probabilities from the football simulation game *Pro Quarterback*. A unique feature of the CSW work was that it used the theory of stochastic games to solve for both the offensive team's and defensive team's optimal strategy mix in each situation. This model simultaneously computed the state values and the fraction of the time the offense and defense should choose each play in a given situation. The CSW work allows us to have values for states involving second and third down while Carter and Machol, Romer, and Footballoutsiders.com only have values for first down. We modified *Pro Quarterback* to include 2006 NFL kickoff return data and 2006 field goal accuracy. Romer used data from the NFL 2001–4 seasons and then solved a complex system of equations to estimate the value of first and 10 for each yard line. Romer's work caused quite a stir because his values led to the inescapable conclusion that teams should go for it on fourth down in many situations in which NFL coaches either punt or try a field goal. For example, when a team is facing fourth and 5 on its own 30-yard line, Romer's values indicate that the team should go for the first down and not punt. Few NFL coaches would go for the first down in this situation. (We will discuss this conundrum in chapter 21.) The fact that we have values for all downs, yard lines, and yards to go situations allows us to evaluate (see chapter 22) the effectiveness of every play run by an NFL team. With values for only first down we cannot evaluate the effectiveness of second- and third-down plays.

Because there are approximately 12,000 possible states and during a typical NFL season fewer than 40,000 plays are run from scrimmage, there are not enough data to estimate the value of every possible state from play-by-play data. That is why we used a board game to estimate the state values.

A Simple Example of State Values

Let's use a simplified example to show how state values can be calculated. Suppose we play football on a 7-yard field, using a template set up with these columns:

My Goal   Yard 1   Yard 2   Yard 3   Yard 4   Yard 5   Opponent Goal

The rules of the game are simple. We have one play to make a first down. It takes 1 yard to get a first down. We have a 50% chance of gaining 1 yard and a 50% chance of gaining 0 yards on any play. When we score, we get 7 points and the other team gets the ball 1 yard from their goal line.

What is the value of each state? We let $V(i)$ be the expected number of points by which we should win an infinite game if we have the ball on yard line i. Then we can use the following equations to solve for $V(1)$, $V(2), \ldots V(5)$:

$$V(1) = .5V(2) - .5V(5) \tag{1}$$

$$V(2) = .5V(3) - .5V(4) \tag{2}$$

$$V(3) = .5V(4) - .5V(3) \tag{3}$$

$$V(4) = .5V(5) - .5V(2) \tag{4}$$

$$V(5) = .5(7 - V(1)) - .5V(1). \tag{5}$$

Recall from chapter 6 the Law of Conditional Expectation tells us that

expected value of random variable

$$= \sum_{\text{all outcomes}} (\text{probability of outcome})$$

$$\times (\text{expected value of random variable given outcome}).$$

To derive equations (1)–(5) we condition whether we gain a yard or not. Suppose we have the ball on the 1-yard line. Then with probability .5 we gain a yard (and the situation is now worth $V(2)$) and with probability .5 we do not gain a yard and the other team gets ball one yard from our goal line. Now the situation is worth $-V(5)$ to us because the other team has the ball at yard 1 and the value to the other team is now equal to what the value would be to us if we had the ball at yard 5. This means that as shown in (1), the value of having the ball on our 1-yard line may be written as $.5V(2) - .5V(5)$. Equations (2)–(4) are derived in a similar fashion.

To derive equation (5) note that with probability .5 we gain a yard and score 7 points. Also, the other team gets the ball on the 5-yard line, which has a value to them of $V(1)$ (and value to us of $-V(1)$). With probability .5 we fail to gain a yard and the other team gets the ball on their 5-yard line, which has a value to them of $V(1)$ (and to us a value of $-V(1)$). Therefore, as shown in (5), the expected value of having the ball on our 5-yard line is $.5(7 - V(1)) - .5V(1)$. Solving these equations, we find that $V(1) = -5.25$, $V(2) = -1.75$, $V(3) = 1.75$, $V(4) = 5.25$ $V(5) = 8.75$.

Thus each yard line closer to our "goal line" is worth 3.5 points $= .5$ touchdown. The trick in adapting our methodology to actual football is that the "transition probabilities" that indicate the chances of going from, say, first and 10 on the 20-yard line to second and 4 on the 26-yard line are difficult to estimate. Despite this difficulty, the next two chapters show that the state value approach provides many insights into effective football decision-making and enables us to evaluate the effectiveness of different types of plays in different situations.

FOOTBALL DECISION-MAKING 101

During the course of a football game, coaches must make many crucial decisions, including the following:

> *1. It is fourth and 4 on the other team's 30-yard line. Should we kick a field goal or go for a first down?*
>
> *2. It is fourth and 4 on our own 30-yard line. Should we go for a first down or punt?*
>
> *3. We gained 7 yards on first down from our own 30-yard line. The defense was offside. Should we accept the penalty?*
>
> *4. On first and 10 from their own 30-yard line our opponent ran up the middle for no gain. They were offside. Should the defense accept the penalty?*
>
> *5. What is the optimal run-pass mix on first down and 10?*

Using the concepts of states and state values discussed in chapter 20, these decisions (and many others) are easy to make. Simply choose the decision that maximizes the expected number of points by which we win a game of infinite length.

Let's analyze the five situations listed above.

1. It is fourth and 4 on the other team's 30-yard line. Should we kick a field goal or go for a first down?

To simplify matters we will assume that if we go for the first down we will get it with probability p (we assume that if we get first down we gain exactly 5 yards) or not get first down with probability $1 - p$. In this case we assume we gain exactly 2 yards. We define V(D [down], YTG [yards to go for a first down], YL [yard line where the team has the ball]) to be the number of points by which we will defeat a team of equal ability from the current point onward in a game of infinite length when we have the ball YL yards from our own goal line (YL = 20 is our 20 and YL = 80 is other team's 20) and it is down D with YTG yards to go for first down. Some examples of the CSW (see chapter 20) values follow:

$$V(1,10,50) = 1.875$$

$$V(3,3,80) = 3.851$$

$$V(2,9,5) = -1.647.$$

Then, after conditioning on outcome of our fourth-down play the expected value of going for the first down is

$$pV(1,10,75) + (1-p)(-V(1,10,28)).$$

If we get the first down we assumed we would gain 5 yards, thus first and 10 on the 75-yard line. If we do not get the first down, then the other team has the ball on their 28-yard line, which has V(1,10,28) to them (or value $-V(1,10,28)$ to us.

Evaluating the Value of a Field Goal Attempt

To evaluate the value of a field goal we need to know how the probability of making a field goal depends on the length of the field goal attempted. Kickoffs are on average returned to the 27-yard line. To simplify our calculations we will assume a kickoff after a field goal is always returned to the 27-yard line. We assume all field goal attempts are made from 7 yards behind the line of scrimmage. For example, if the line of scrimmage is the 30-yard line, then the kick is attempted from 37-yard line and is 47 yards long. If the kick is missed, the defending team will get the ball on their own 37-yard line. Table 21.1 shows the field goal accuracy for the 2006 NFL season.

We need to use the data in the table to determine how the probability of making a field goal depends on the length of the kick. If we assume that the field goals listed in table 21.1 are kicked from the 25-, 35-, 45-, and 53-yard lines, respectively, then, as shown in figure 21.1, the probability of making a kick is not a straight-line function of the length of the kick. Therefore it is unreasonable to try and fit a straight-line relationship of the form

probability kick is good = a + b(length of kick).

Statisticians have found that using a straight line to estimate how an independent variable influences the probability of an event happening usually

TABLE 21.1
Field Goal Accuracy as a Function of Distance, 2006

Length of Field Goal	Made	Attempted	Percentage Made
20–29 yards	252	264	95.1
30–39 yards	232	268	86.9
40–49 yards	211	283	74.5
> 50 yards	39	81	48.1

Source: http://www.nfl.com.

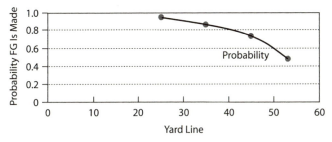

Figure 21.1. Field goal accuracy as function of yard-line position.

gives poor results. Statisticians have found, however, that an equation of the form

$$\text{Ln}\left(\frac{p}{1-p}\right) = a + b(\text{length of kick})$$

does a good job of estimating how the probability p of making a field goal depends on the length of the kick. (The file Fg.xls on our Web site details how we fit this equation to the data in table 21.1.) The above equation is an example of logistic regression. Logistic regression postulates that $\text{Ln}\left(\frac{p}{1-p}\right)$ is a linear function of one or more independent variables. In this case the single independent variable is the length of the field goal attempt. Logistic regression is the method generally used to estimate how the probability of an event depends on one or more independent variables.

We found a = 7.05 and b = −.134. This yields the probability of making field goals shown in table 21.2. Now our table of values shows that V(1,10,75) = 3.884, V(1,10,28) = .336, V(1,10,37) = .979, and V(1,10,27) = .266. Therefore, the expected value by which we will win a game of infinite length if we go for it on fourth down is 3.884p + (1 − p)(−.336).

Note that the probability of making field goal is .676 (because attempt is 47 yards). Since we must kick off after a made field goal, the expected value of a successful field goal is 3 − V(1,10,27). The expected value by which we win a game of infinite length if we attempt a field goal is

$$(.676)(3 − V(1,10,27)) + (1 − .676)(−V(1,10,37))$$
$$= .676(2.734) + (.324) \times (−.979) = 1.531.$$

Thus if 3.884p + (1 − p)(−.336) ≥ 1.531, we should go for a first down. Solving this inequality, we find we should go for the first down if the probability of making the first down is at least 44%. Alamar tabulated the number of yards gained on each running and passing play on first down and 10 during the 2005 season. He found that 43% of all runs on first and 10 gained at least 4 yards and 55% of all pass attempts gained 4 or more yards. It is likely more difficult to make 4 yards on fourth and 4 than on first and 10, so it appears that going for it on fourth and 4 is a close call. Table 21.3 shows the probability for successful first-down conversion for each situation, based on data from the 2002–5 NFL seasons.[1] Since the probability of making a first down on third or fourth down and 4 yards to go exceeds 44%, we should go for it on fourth down.

2. It is fourth and 4 on our own 30-yard line. Should we go for a first down or punt?

Let's assume that if we get the first down we gain 5 yards and we now have value V(1,10,35) = .839. If we fail to get the first down we assume a gain of 2 yards. Now the other team has the ball on the 68-yard line, which is worth −V(1,10,68) = −3.265 to us. Assume that if we punt we will always net 40 yards (during the 2007 NFL season the average net gain on a punt was 39.1 yards), which puts us in a situation worth −V(1,10,30) = −.48 points to us. Therefore we should go for the first down if probability p of obtaining first down satisfies (.839)p − 3.265(1 − p) ≥ −.48. This inequality is satisfied for p ≥ .678. Thus we would need a 67.8% chance of success to justify going for

[1] Alamar, "The Passing Premium Puzzle"; http://www.pro-football-reference.com/blog/?p=50.

TABLE 21.2
Estimated Probabilities of Field Goal Success

Yard Line	Prob FG good	Yard Line	Prob FG good
18	0.990343545	38	0.874819179
19	0.988970783	39	0.859355781
20	0.987405351	40	0.842326493
21	0.985620959	41	0.823658399
22	0.98358796	42	0.803296202
23	0.981272984	43	0.781207347
24	0.978638563	44	0.757387178
25	0.975642746	45	0.7318638
26	0.9722387	46	0.704702208
27	0.968374343	47	0.676007244
28	0.963991992	48	0.645924914
29	0.959028075	49	0.614641689
30	0.953412923	50	0.582381532
31	0.947070692	51	0.549400598
32	0.939919458	52	0.515979754
33	0.931871532	53	0.482415361
34	0.922834072	54	0.4490089
35	0.912710062	55	0.41605622
36	0.901399719	56	0.383837206
37	0.888802418	57	0.352606624

the first down. Alamar's data and Pro-football-reference.com indicate that in this situation we should certainly (as NFL coaches do) punt.

As pointed out by Phil Birnbaum,[2] Romer's research indicates that a team should go for the first down on its own 30-yard line if it has (approximately) at least a 45% chance of success. This hardly seems reasonable.

[2] See http://sabermetricresearch.blogspot.com/2007_01_01_archive.html.

TABLE 21.3
Probability of Successful Third- or Fourth-
Down Conversion

Yards to Go	Probability Third- or Fourth-Down Play Makes the First Down
1	.67
2	.52
3	.53
4	.48
5	.41

Source: http://www.pro-football-reference.com/
blog/?p=50.

Romer's work also indicates that even on its own 10-yard line, a team should go for it on fourth and 3. Our work suggests that going for it on fourth and 3 from a team's own 10-yard line requires at least a 71% chance of a successful fourth-down conversion.

3. We gained 7 yards on first down from our own 30-yard line. The defense was offside. Should we accept the penalty?

After the play we have $V(2,3,37) = 0.956$. If we accept the penalty we have $V(1,5,35) = 0.983$. Therefore, we should accept the penalty. If we gained 8 yards we would have $V(2,2,38) = 1.068$ points and we should accept the play. Thus, gaining 8 or more yards on first down is better than accepting a 5-yard first-down penalty.

4. On first and 10 from their own 30-yard line our opponent ran up the middle for no gain. They were offside. Should we accept the penalty?

After the run our opponent has $V(2,10,30) = .115$. If we accept the penalty our opponent has $V(1,15,25) = -.057$, so we should accept the penalty. Since $V(2,11,29) = -.007$ and $V(2,12,28) = -.125$, it appears that the defense should decline a first-down 5-yard penalty if a first-down play loses 2 or more yards.

5. What is the optimal run-pass mix on first down and 10?

Alamar showed that during the 2005 season NFL teams ran 50.2% of the time and passed 49.8% of the time. Alamar defines a play on 1st and 10

TABLE 21.4
Distribution of Yards Gained on Passing and Running Plays

Yards Gained	Frequency	Interceptions
−6	161	
0	801	24.831
1	38	1.178
2	42	1.302
3	58	1.798
4	84	2.604
5	99	3.069
6	106	3.286
7	97	3.007
8	90	2.79
9	107	3.317
10	34	1.054
13	302	9.362
18	149	4.619
25	100	3.1
44	65	2.015
66	22	0.682
76	5	0.155

as a success if it gains at least 4 yards. Many football data tabulation software programs (such as Pinnacle Systems) agree with this definition. Alamar then points out that only 42% of running plays are successful on first and 10, while 53.5% of all passing plays are successful on first and 10. Since passes are more successful than runs on first down, Alamar believes this indicates that teams should pass more on first down.[3] I agree with this assessment, but by using our state values we can quantify how much better it is to pass than run on first and 10. Table 21.4 shows the distribution of yards

[3] Alamar, "The Passing Premium Puzzle."

gained on running and passing plays on first and 10. Alamar does not give yards gained on a rush that lost yardage. I assumed that all rushes that lost yardage lost 2 yards and that all passes that resulted in quarterback sacks were for losses of 6 yards. In addition, 2.1% of all rushes led to lost fumbles and 3.1% of all passes led to interceptions, so I factored these percentages into the calculations and assumed the location of fumbles and interceptions were distributed identically to the distribution of yards gained rushing and passing on plays that did not result in a turnover or a sack. Finally, I assumed all rushes that were listed as gaining between 20 and 30 yards gained 25 yards, and so forth.

We assume that the team starts first and 10 on their own 20-yard line. Each running attempt added an average of -0.044 points while the average passing attempt added 0.222 points per attempt. Thus, running on first down is much less effective than passing. Therefore, teams should pass more often on first down. Of course, the success of a given play depends on the defensive team's setup. This means a team cannot pass all the time because then the defense will anticipate the pass and always play defense for a pass attempt. This is why football teams practice the mixed strategies that are the cornerstone of two-person zero sum game theory, which will be discussed in chapter 23.

22

A STATE AND VALUE ANALYSIS OF THE 2006 SUPER BOWL CHAMPION COLTS

By winning the Super Bowl, the 2006 Indianapolis Colts brought great joy to the Hoosier state. In this chapter we will use the state and value approach described in chapter 20 to answer many interesting questions about the Colts' offense, such as the following:

1. *On any down and yards to go situation, is running more or less effective than passing? For example, on first and 10, is running more effective than passing?*
2. *Are runs more or less effective than passes overall?*
3. *Was Joseph Addai a more effective runner than Dominique Rhodes?*
4. *Who is better: Marvin Harrison or Reggie Wayne?*
5. *Is it better to throw deep or short?*
6. *Are the Colts more effective running right, left, or up the middle?*

ESPN.com and NFL.com give complete play-by-play logs for each NFL game. For each play we are given the following information:

- the down before the play
- yards to go for a first down before the play
- the yard line before the play
- whether the play was a run or pass
- the ball carrier on a running play
- behind which offensive line position (left end, left tackle, left guard, middle, right guard, right tackle, right end) the carrier ran on a running play
- the quarterback and intended receiver on a passing play
- whether the pass was deep or short
- the final result of the play, down, and yards to go for a first down and new yard line after play is completed

For example, here is the entry for the key play in the 2008 Super Bowl:

	I	J	K	L	M	N	O	P	Q	R
8	Play Number	Start Down	Start Yards to Go	Start Deadline	End Down	End Yards to Go	End Yardline	Start Value	End Value	Delta
9	1	1	10	34	1	10	47	0.769	1.67	0.898
10	2	1	10	47	2	6	51	1.667	1.78	0.113
11	3	2	6	51	3	6	51	1.78	1.25	−0.53
12	4	3	6	51	1	10	58	1.25	2.47	1.215
13	5	1	10	58	2	2	66	2.465	3.19	0.725
14	6	2	2	66	3	1	67	3.19	2.92	−0.273
15	7	3	1	67	1	10	69	2.917	3.35	0.43
16	8	1	10	69	2	14	65	3.347	2.13	−1.219
17	9	2	14	65	3	13	66	2.128	1.5	−0.63
18	10	3	13	66	1	10	80	1.498	4.37	2.872

Figure 22.1. Play-by-play data for the Colts, 2006: offense.
Note: Every offensive play for the Colts' 2006 season was entered into the file Val2727.xlsx.
Source: http://www.espn.com.

3rd and 5 at NYG 44 (1:15) (Shotgun) E. Manning pass deep middle to D. Tyree to NE 24 for 32 yards (R. Harrison).

Using our chart of values from the file Val2727.xlsx for each state we can determine the "point value added" by each play. We will use the concepts of states and values developed in chapters 20 and 21 to answer the questions posed above about the Colts' offense. Figure 22.1 analyzes the net benefits (in points) "earned" by each play from scrimmage.

We see that the first play began with first down and 10 yards to go on the Colts' 34-yard line. Peyton Manning threw a short pass to Reggie Wayne, which gained 13 yards. Now the Colts had first down and 10 on their own 47-yard line. At the start of the play the value of the situation was 0.769 points and after the play the value of the situation was 1.667 points. Therefore this play generated 0.898 points of value.

On the second play Dominique Rhodes ran around right end and gained 4 yards. This play generated 0.113 points of value. Using the Excel 2007 AVERAGEIFS, SUMIFS, and COUNTIFS functions (see the chapter appendix), we can summarize these data in many ways and gain valuable insight into what makes the Colts' offense tick.

8	Play Number	Run or Pass?	Runner	Run or Pass Location	Quarterback	Receiver
9	1	P		Short	P. Manning	R. Wayne
10	2	R	D. Rhodes	RE		
11	3	P		Short	P. Manning	R. Wayne
12	4	P		Short	P. Manning	B. Utrecht
13	5	P		Short	P. Manning	M. Harrison
14	6	R	D. Rhodes	LT		
15	7	R	D. Rhodes	LT		
16	8	P			P. Manning	
17	9	R	D. Rhodes	RT		
18	10	P		Short	P. Manning	D. Rhodes

Figure 22.1. (*cont.*)

1. On first and 10, is running more effective than passing?

The Colts ran 222 times on first and 10 and averaged 0.119 points per run. They passed 204 times on first and 10 and averaged 0.451 points per pass. This shows that for the Colts passing on first down is much more effective than running. Given the average quality of defense play the Colts see on first and 10, it appears that changing a single run into a pass would on average generate $.451 - .119 = .332$ points. Thus if the Colts passed more (and the defensive team did not change their mix of defensive calls), they would benefit by passing slightly more than 48% of the time on first and 10. If the Colts passed much more often than their current level, the defense would probably call more plays geared to stop the pass. The mean value of 0.451 points per pass would probably drop, and the mean value of 0.119 points per run would probably increase.

On second down with 5 to 10 yards to go for a first down, running was more effective. The Colts averaged 0.240 points per running play and 0.218 points per passing play.

2. Are runs more or less effective than passes overall?

On all passing plays the Colts averaged 0.416 points per pass and on all running plays they averaged 0.102 points per run. The Colts currently pass 58% of the time, but this analysis indicates that they should pass even more.

3. Was Joseph Addai a more effective runner than Dominique Rhodes?

On Joseph Addai's running plays the Colts averaged 0.134 points per run, and on Dominique Rhodes's running plays the Colts averaged 0.041 points per run. Thus, Joseph Addai generated roughly 0.09 more points per carry than Rhodes. The Colts let Rhodes go to the Raiders after the 2006 season because they were confident that Addai was a better runner.

4. Who is better, Marvin Harrison or Reggie Wayne?

Figure 22.2 illustrates the points generated per pass to Marvin Harrison and Reggie Wayne. Passes to Wayne generated nearly 0.11 points more per pass than passes thrown to Harrison. Of course, this could be because Harrison was being double teamed, which set up Wayne. Also note that passes to running back Addai were much more effective than passes to tight end Dallas Clark and passes to running back Dominique Rhodes were the least effective.

	L	M	N	O	AE	AF
1	**Run or Pass?**	**Player**	**Count**	**Average**	**Count**	**Average**
2	R	J. Addai	223	0.133767	51	0.458824
3	R	D. Rhodes	184	0.040804	46	0.197804
4	P	M. Harrison	152	0.563217		
5	P	R. Wayne	140	0.672493		
6	P	D. Clark	58	0.225414		
7	P	B. Utrecht	57	0.22386		
8	P	B. Fletcher	25	0.5586		

Figure 22.2. Colts' receivers' effectiveness, 2006. From Figureval2727.xls.

5. Is it better to throw deep or short?

Twenty-two percent of the Colts' passes were thrown deep and 78% were thrown short. Deep passes averaged 0.951 points per attempt while short passes averaged 0.318 points per attempt (less than one-third as much as a deep pass). This indicates that the Colts should try many more long passes.

6. Are the Colts more effective running right, left, or up the middle?

Figure 22.3 shows an analysis of the Colts' running plays broken down by the location of the run. The Colts are most effective running behind Pro Bowl left tackle Tarik Glenn. They are also effective running wide around left end. Running off left guard or right end is relatively ineffective.

	Y	Z	AA
17	Run Location	Frequency	Mean Delta
18	LE	66	0.22601515
19	LT	48	0.30933333
20	LG	25	−0.20172
21	MID	124	0.1195
22	RG	32	0.1219375
23	RT	46	0.08178261
24	RE	78	−0.05564103

Figure 22.3. Colts' running effectiveness by location, 2006.

This brief analysis of the Colts' offense has yielded many interesting results. Using the state value approach to further analyze our data would probably yield many more interesting results.

Footballoutsiders.com Brings Sabermetrics to the NFL

Aaron Schatz and his colleagues at Footballoutsiders.com have done a great job using state and value concepts and sabermetric ideas to rate NFL players and teams. To rate a quarterback, for example, Schatz and his colleagues look at every play involving the quarterback. They have a measure of the effectiveness of each play that is similar to the change in state value we used in our analysis of the Colts. Then they compare the quarterback's average effectiveness to the average effectiveness of all NFL passing plays in similar down, yard-line, and yards to go situations. Comparing the quarterback average to the overall league quarterback average and adjusting for the strength of the defensive teams faced by the quarterback yields the measure DVOA (Defense Adjusted Value over Average). For example, during the Patriots 2007 regular season, Tom Brady's DVOA was 62%. This indicates that Brady played 62% better than an average quarterback (of course, much of his effectiveness resulted from the play of his great receivers and offensive line). Carolina Panthers quarterback David Carr, on the other hand, played 32.5% below average. Schatz and his colleagues have also adapted the sabermetrics concept of a replacement player to create another measure of player effectiveness, called DPAR (Defense Adjusted Points above Replacement). During 2007, for example, Tom Brady had a DPAR rating of 200.2 points, which indicated that Brady contributed 200.2 points more value than did a "replacement quarterback."

APPENDIX

Use of COUNTIFS, AVERAGEIFS, and SUMIFS Functions in Excel 2007

Our analysis of the Colts 2006 play-by-play data is greatly simplified by the fact that Excel 2007 includes the powerful COUNTIFS, AVERAGEIFS, and SUMIFS functions. This section includes several examples of how these functions work.

In the cell range AA2:AA8 I used the COUNTIFS function to determine how many running or passing plays involved a given player. For example, in cell AA2 (see figure 22.4) the formula

$$=COUNTIFS(\$S\$9:\$S\$1016,Y2,\$T\$9:\$T\$1016,Z2)$$

counts the number of plays that have an R in column S (for run) and a JA in column T (for Joseph Addai). Therefore, this formula counts how many times Joseph Addai carried the ball. We find Addai carried the ball 223 times.

In cell AB2, the formula

$$=AVERAGEIFS(\$R\$9:\$R\$1016,\$S\$9: \\ \$S\$1016,Y2,\$T\$9:\$T\$1016,Z2)$$

averages the delta for each play (the delta for each play is in column R) for which the entry in column S was an R (for run) and the entry in column T was JA (for Joseph Addai). The formula in cell AB2 calculates that the average change in point "value" when Addai ran the ball was 0.134.

We can also use the SUMIF and COUNTIFS functions to do our calculations (see figure 22.5). For example, in cell AA19 the formula

$$=SUMIF(\$U\$9:\$U\$1016,Y19,\$R\$9:\$R\$1016)/Z19$$

	Y	Z	AA	AB
1	Play	Player	Count	Average
2	R	J. Addai	223	0.133767
3	R	D. Rhodes	184	0.040804
4	P	M. Harrison	152	0.563217
5	P	R. Wayne	140	0.672493
6	P	D. Clark	58	0.225414
7	P	B. Utrecht	57	0.22386
8	P	B. Fletcher	25	0.5586

Figure 22.4. Colts' analysis of plays, 2006.

	Y	Z	AA
17	**Run**	**Frequency**	**Mean Delta**
18	LE	66	0.226015152
19	LT	48	0.309333333
20	LG	25	−0.20172
21	MID	124	0.1195
22	RG	32	0.1219375
23	RT	46	0.081782609
24	RE	78	−0.05564103

Figure 22.5. Colts' average delta for running plays by location, 2006.

computes that when the Colts ran off left tackle, on average they improved their point "value" by nearly 0.31 points per run. Cell Z19 uses the COUNTIFS function to track how many times the Colts ran off left tackle. Then the SUMIF function adds up the deltas in column R for each play having a left tackle in column U. Of course, when we want to sum numbers involving one column based on criteria that involve more than one condition, we can use the SUMIFS function. This function works just like the AVERAGEIFS function, but it computes a sum rather than an average.

23

IF PASSING IS BETTER THAN RUNNING, WHY DON'T TEAMS ALWAYS PASS?

In football the offense selects a play and the defense lines up in a defensive formation. Let's consider a very simple model of play selection in which the offense and defense simultaneously select their play:

- The offense may choose to run or pass.
- The defense may choose a run or pass defense.

The number of yards gained is given in table 23.1, which we call a payoff matrix for the game. We see that if the defense makes the right call on a run, the opposing team loses 5 yards, and if the defense makes the wrong call, the team gains 5 yards. On a pass the right defensive call results in an incomplete pass for the opposing team while the wrong defensive call results in a 10-yard gain for the team. Games in which two players are in total conflict are called two-person zero sum games (TPZSGs). In our game every yard gained by the offense makes the defense one yard worse-off, so we have a TPZSG. The great mathematician James von Neumann and the brilliant economist Oskar Morgenstern discovered the solution concepts for TPZSG.[1] We assume the row player wishes to maximize the payoff from the payoff matrix and the column player wants to minimize the payoff from the payoff matrix. We define the value of the game (v) to the row player as the maximum expected payoff the row player can assure himself. Suppose we choose a running play. Then the defense can choose run defense and we lose 5 yards. Suppose we choose a pass offense. Then the defense can choose pass defense and we gain 0 yards. Thus by throwing a pass the offense can ensure themselves of gaining 0 yards. Is there a way the offense can assure that on average they will gain more than 0 yards? A

---

[1] Von Neumann and Morgenstern, *Theory of Games and Economic Behavior*.

TABLE 23.1
Payoff Matrix for Football Game (Yards)

	Run Defense	Pass Defense
Offense runs	-5	5
Offense passes	10	0

player in a TPZSG can choose to play a mixed strategy in which he makes choices with a given probability. Let's suppose the offense chooses run with probability q and chooses pass with probability $1 - q$. On average, how does our mixed strategy do against each of the defense's "pure" strategy choices?

- If a run defense is chosen, the expected gain is $q(-5) + (1-q)10 = 10 - 15q$.
- If a pass defense is chosen, the expected gain is $q(5) + (1-q)(0) = 5q$.

For any value of q chosen by the offense, the defense will choose the defense that yields minimum$(10 - 15q, 5q)$. Thus the offense should choose the value of q between 0 and 1 that maximizes minimum$(10 - 15q, 5q)$. From figure 23.1 we see that this minimum is attained where $10 - 15q = 5q$ or $q = 1/2$.

Thus the optimal mixed strategy for the offense is to run half of the time and pass half of the time. This ensures the offense of an expected payoff of at least 5/2 yards. By choosing $q = 1/2$ the offense is assured an expected gain of 5/2 yards. This makes the game very dull. No matter what strategy (pure or mixed) the defense chooses, the gain is on average 5/2 yards. Therefore, the value of this simple game to the row player is 5/2 yards.

Note that when the defense guesses correctly, passing is 5 yards better than running, and if the defense guesses incorrectly, passing is 5 yards better than running (see table 23.1). Despite the fact that passing seems like a much better option, our optimal strategy is to run half of the time. Note that TPZSG theory provides a rationale for the fact that teams actually do "mix up" running and passing calls. For similar reasons, players should occasionally bluff in poker when they have a poor hand. If you never bet with a poor hand, your opponents may not call your bet very often when you have a good hand. Bluffing with a bad hand will also sometimes enable you to win a hand with poor cards.

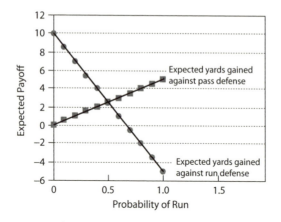

Figure 23.1. Offense's optimal mix.

Now let's find the defensive team's optimal strategy. Let x = the probability that the defense calls a run defense and 1 − x = probability that defense calls a pass defense. The defense's goal is, of course, to minimize the expected payoff to the offense. Against this defensive mixed strategy, how will the offense fare?

- If offense chooses run, then on average they gain $x(-5) + (1 - x)(5) = 5 - 10x$.
- If offense chooses pass, then on average they gain $x(10) + (1 - x)(0) = 10x$.

Therefore, for any value x chosen by the defense, the offense will choose the play attaining maximum$(5 - 10x, 10x)$. The defense should therefore choose the value of x that minimizes maximum$(5 - 10x, 10x)$. Figure 23.2 shows that choosing $x = 1/4$ will attain this minimum and ensure that the offense is held to an average of 5/2 yards per carry. Therefore, the defense should play a pass defense 75% of the time. This is because the pass play is better than the run. Since the defense looks for the pass more often, the offense only passes half the time, even though the pass seems like a much more effective play.

Note that we found the offense can ensure itself of and the defense can hold the offense (on average) to the same number: 5/2 yards; 2.5 yards is the value of the game. Von Neumann and Morgenstern proved that the values obtained by analyzing the row and column players' strategies in a TPZSG are always identical.

In general suppose our payoff matrix looks like that in table 23.2. If we assumed that the defense would choose a run and defense with equal

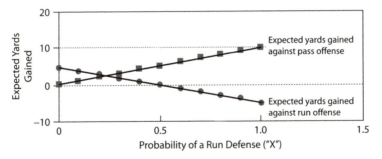

Figure 23.2. Defense's optimal mix.

TABLE 23.2
Generalized Football Payoff Matrix

	Run Defense	Pass Defense
Offense runs	$r - k$	$r + k$
Offense passes	$p + mk$	$p - mk$

probability, then we can interpret this matrix as indicating that running plays gain r on average and passing plays gain p on average. Also the correct choice of defense appears to have m times as much effect on a passing play as it does on a running play. For example, if $m = 2$, then for a pass a correct defense makes the pass gain 4k yards worse than the incorrect defensive choice. For a run, the correct choice of defense makes the run gain 2k yards worse than the incorrect defensive choice.

Note that our previous example shows that for a TPZSG with a 2×2 payoff table, the optimal strategy for a player is found by equating the expected payoff against both of the opponent's strategies. For the payoff matrix shown in table 23.2, we can show that the optimal strategy for the offense is to run with probability $m/(m + 1)$ and pass with probability $1/(m + 1)$. The optimal defensive mix can be shown to be choosing a run defense with probability $.5 + \dfrac{r - p}{2k(m + 1)}$ and a pass defense with probability $.5 - \dfrac{r - p}{2k(m + 1)}$. Note that the defense chooses to defend the better play more than 50% of the time. For $m = 2$ the offense will run 2/3 of the time and pass 1/3 of the time. For $m = 0.5$ it runs 1/3 of the time and passes 2/3 of the time. For m = 1 it runs and passes with equal probability. The idea is that for m >1 the defensive call has more effect on a pass than a run, and for m < 1 the defensive call has more effect on a run than a pass. Thus we see that the offense will choose more often the play over which the defense has "less control." For m > 1 the defense has less control over the run, so it runs more than it passes. Similarly, if m < 1 the defense has less control over the pass, so the offense passes more than it runs. Note the optimal run-pass mix does not depend on r and p, which represent the base effectiveness levels of running and passing plays, respectively.

We can draw several other interesting insights from this simple formulation.

- Suppose we get a new and improved quarterback. We can model this by saying that for any defense our pass play will gain, say, 3 more yards. Should we pass more often? We are simply replacing the current value of p by p + 3 and leaving m unchanged, so our optimal run-pass mix remains unchanged. Even though passing has improved, the defense will play a pass defense more often, so we should play the same run-pass mix as before.
- Suppose we get a new and improved running back who can gain 5 more yards per carry against a pass defense but no more against a run defense. How will the optimal run-pass mix change? Intuitively one would think the offense should run more often because of the improved ground game.

Table 23.3 illustrates that this is not the case.

The revised payoff matrix is shown in table 23.4. The optimal run-pass mix is to run 2/5 of the time and pass 3/5 of the time. The value of the game

TABLE 23.3
Original Payoff Matrix for Football Paradox (Yards)

	Run Defense	Pass Defense
Offense runs	−5	5
Offense passes	5	−5

Note: This matrix has r = 0, k = 5, p = 0, and m = 1. Therefore, the optimal strategy is to run 50% of the time and pass 50% of the time. The value of this game is 0 yards. The optimal defensive mix is to choose a run defense half of the time and a pass defense half of the time.

Source: Mike Shor, "Game Theory and Business Strategy Mixed Strategies in American Football," http://www2.owen.vanderbilt.edu/Mike.Shor/courses/game-theory/docs/lecture05/Football.html.

TABLE 23.4
Revised Payoff Matrix for Football Paradox (Yards)

	Run Defense	Pass Defense
Offense runs	−5	10
Offense passes	5	−5

Source: Mike Shor, "Game Theory and Business Strategy: Mixed Strategies in American Football," http://www2.owen.vanderbilt.edu/Mike.Shor/courses/game-theory/docs/lecture05/Football.html

has increased to 1 yard and the team runs less often. The optimal run-pass defense mix is run defense 3/5 and pass defense 2/5. Since the offense's running game has improved so much against the pass defense, the defense is terrified of the run and plays a run defense more often. This causes the offense to pass more and run less, even though its running game is much better.

Game Theory and Real Football

The TPZSG examples in this chapter are admittedly simple and bear little relationship to actual football. Is there a way an NFL team can actually use TPZSG to create an optimal play mix? For game theory to be useful in the NFL we need the following information for each play:

- play called by offense
- defensive formation or strategy
- down, line of scrimmage, and yards gained on play

First we combine the down and yards to go situation into groups that might look like this:

- first and 10
- second and short (≤ 3 yards for first down)
- second and medium (4 – 7 yards for first down)
- second and long (at least 8 yards to go for a first down)
- third and short (≤ 2 yards for a first down)
- third and medium (3–5 yards for a first down)
- third and long (more than 5 yards for a first down)

Suppose the offense has fifteen plays they can call on first and 10 and the defense has ten plays. Given the results of each play we could determine for a given down and yards to go situation the average value points gained on plays for each offensive and defensive play call combination. For example, we might find that on first and 10 LaDainian Tomlinson sweeping right end for the San Diego Chargers averaged 0.4 value points against the cover two defense. This results in one of the 150 entries in the relevant payoff matrix for first and 10. Then we can use TPZSG to solve for the optimal mixed strategies for both the offense and defense. For example, we might find on first and 10 the optimal offensive play selection mixed strategy would ensure (against any defensive mix) an expected value of 0.3 points. The numbers in the matrix could be adjusted based on the strength or weakness of the opposition. To get to this game theory "Nirvana" we would need to have coaching experts break down the film of every play to tell us the offensive and defensive play calls.

I am confident that once this data collection hurdle is overcome, game theory will provide useful insights into football play selection.

24

SHOULD WE GO FOR A ONE-POINT OR TWO-POINT CONVERSION?

Since 1994, when the NFL began allowing teams to go for a two-point conversion after a touchdown, it has become important for NFL coaches to determine whether to go for one or two points after a touchdown. The success rate for a one-point conversion is over 99%, so we will assume that there is a 100% chance that a one-point conversion will be successful. The success rate for two-point conversions is around 47%.[1] On average, a one-point conversion try earns one point and a two-point conversion attempt earns $0(.6) + 2(.47) = .94$ points. So, on average, a one-point conversion earns more points but in some situations it is clear that going for two is the proper choice. For example, if a team scores a touchdown with thirty seconds to go and they were down by eight points before the touchdown, the team needs to go for two to tie the game. Most coaches have a "chart" that tells them whether they should go for one or two points based on the score of the game. The idea of the "chart" is believed to have originated with UCLA assistant coach Dick Vermeil during the early 1970s.[2]

The coach's decision should depend on the amount of time left in the game as well as the score. For example, if there is a lot of time remaining in the game, then if the team scores a touchdown and is down by eight points they may not want to choose a play (the two-point conversion) over a play with higher expected scoring (the one-point conversion). To determine how the optimal strategy depends on the score of the game and time remaining we need to use a sophisticated technique, dynamic programming, which allows us to work backward from the end of the game to-

---

[1] Schatz, *Football Prospectus*.

[2] Vermeil later became a successful NFL coach and NFL TV commentator. Sackrowitz's "Refining the Point(s)-After-Touchdown Scenario" appears to be the first mathematical study of the one- or two-point conversion decision.

ward the beginning of the game. Dynamic programming was developed during the 1950s by Richard Bellman.[3] Using statistics from the 2006 season, we assume that on any possession there is a .19 chance of scoring a touchdown, a .13 chance of scoring a field goal, and a .68 chance of not scoring. We assume two equally matched teams are playing (our model can easily handle teams of different abilities). We ignore the possibility of a safety or that the defensive team scores on a possession. The less mathematically sophisticated reader may (without loss of continuity) now skip to the section of this chapter titled "The Chart."

We define $F_n(i)$ = probability the team wins the game if they are i points ahead, they have just gotten the ball, and n possessions remain. We define $G_n(i)$ = probability the team wins the game if they are i points ahead, their opponent has just gotten the ball, and n possessions remain. N = 0 means the game has ended. If the game is tied, we assume each team has a 0.50 chance of winning in overtime. We will assume that i is always between -30 and $+30$ points inclusive. (The calculations are in the file two-points.xls.) Although past data assume that on average, 47% of two-point conversions are successful, we will proceed pessimistically and assume only 42% of all two-point conversion attempts succeed. In our spreadsheet we can change the chance of a successful two-point conversion, if desired. Clearly

$$G_0(i) = F_0(i) = 1 \text{ for } i > 0, G_0(i) = F_0(i) = 0 \text{ for } i < 0,$$
$$\text{and } G_0(0) = F_0(0) = .5.$$

Now that we know what happens when the game is over we can work backward and determine what happens with one possession left.

$$F_1(i) = .13G_0(\min(i + 3, 30))$$
$$+ .19\max(G_0(\min(i + 7, 30)), .42G_0(\min(i + 8, 30))) \qquad (1)$$
$$+ .58G_0(\min(i + 6, 30)) + .68G_0(i)).$$

$$G_1(i) = .13F_0(\max(i - 3, -30))$$
$$+ .19\min(F_0(\max(i - 7, -30)), .42F_0(\max(i - 8, -30)) \qquad (2)$$
$$+ .58F_0(\max(i - 6, -30)) + .68F_0(i)).$$

Equation (1) uses the Law of Conditional Expectation (see chapter 6). We simply multiply the probability of each possible outcome on the game's

---

[3] Richard Bellman, *Dynamic Programming* (Princeton University Press, 1957).

last possession (field goal, touchdown, or no score) by the probability of winning given each possible outcome of the possession.

- With probability .13 the team kicks a field goal and is now ahead by $\min(i + 3,30)$ points. They win the game with probability $G_0(\min(i + 3,30))$. (Note we truncate the maximum amount the team can be ahead at 30 points.)
- With probability .19 the team scores a touchdown and is now ahead by $\min(i + 7,30)$ points. They win the game with probability $\max(G_0(\min(i + 7,30)), .42G_0(\min(i + 8,30)) + .58G_0(\min(i + 6,30)))$.

(This assumes the team chooses the strategy [go for 1 or go for 2] that maximizes their chance of winning the game.)

- With probability .68 the team does not score. Their probability of winning the game is now $G_0(i)$.

Equation (2) is derived in a similar fashion. Note that the opposition chooses the strategy (one-point or two-point attempt) that minimizes the first team's chance of winning the game. Again we truncate the maximum score differential in the game at 30 points.

Note that if the max in (1) is attained by choosing a one-point conversion, the team should go for one point with one possession left, and if the max in (1) is attained by choosing a two-point conversion, the team should go for two points.

Once we have determined $F_n(i)$ and $G_n(i)$ for $i = -30, -29, \ldots, 39, 30$ we can compute the probability of winning with $n + 1$ possessions remaining using the following recursions:

$$
\begin{aligned}
F_{n+1}(i) = {} & .13G_n(\min(i + 3,30)) \\
& + .19\max(G_n(\min(i + 7,30)), .42G_n(\min(i + 8,30)) \qquad (3) \\
& + .58G_n(\min(i + 6,30)) + .68G_n(i).
\end{aligned}
$$

$$
\begin{aligned}
G_{n+1}(i) = {} & .13F_n(\max(i - 3,-30)) \\
& + .19\min(F_n(\max(i - 7,-30), .42F_n(\max(i - 8,-30)) \qquad (4) \\
& + .58F_n(\max(i - 6,-30)) + .68F_n(i).
\end{aligned}
$$

Again the optimal strategy for the team is to choose a one-point conversion when they have the ball with $n + 1$ possessions remaining if going for one point attains the maximum in (3). Otherwise, they choose a two-point conversion try.

We terminated our calculations with twenty-five possessions remaining the game. On average, a possession consumes 2.6 minutes,[4] so we would expect on average twenty-three possessions per game.

The "Chart"

Table 24.1 summarizes a team's optimal conversion attempt strategy based on the number of points the team is ahead. We can determine approximately how many possessions remain by dividing the time left by 2.6 (the average number of minutes in a possession). Thus, at the beginning of the fourth quarter we can assume n = 5 or 6. We note that in the −22, −19, −16, −11, −8, −5, and −1 scenarios NFL coaches try a two-point conversion over 50% of the time.[5] For all these scores table 24.1 indicates that the team should go for two points near the end of the game. Also, NFL coaches try for two points less than 5% of the time in the first three quarters and try for two points 15% of the time in the fourth quarter. Again the logic of trying more two-point conversions near the end of the game is supported by our chart.

TABLE 24.1
The Chart

Point Differential at Decision	Go for Two Points if Fewer Than This Many Possessions Remain
−30	<=13
−29	<=17
−28	<=18
−27	never
−26	never
−25	<=22
−24	<=8
−23	<=15
−22	<=8
−21	<=9

---

[4] See the Drive Statistics in Schatz, *Football Prospectus*.
[5] See Schatz, *Football Prospectus*.

TABLE 24.1 (Continued)

Point Differential at Decision	Go for Two Points if Fewer Than This Many Possessions Remain
−20	never
−19	<=14
−18	<=19
−17	never
−16	<=6
−15	<=16
−14	<=12
−13	never
−12	never
−11	<=10
−10	never
−9	never
−8	<=20
−7	never
−6	never
−5	<=11
−4	>=5
−3	never
−2	never
−1	<=14
0–5	never
6	<=17
7–12	never
13	<=13
14–30	never

Intuition behind the Chart

Some of these results are intuitively obvious. For example, suppose it is relatively late in the game and a team is down one point and scores a touchdown. They are now up five points. Going for two points and succeeding ensures that a touchdown will not put the team behind, while if they go for one point, a touchdown will put them behind.

Some of the results are counterintuitive. Suppose a team is down fourteen points late in the game and scores a touchdown. Virtually all NFL coaches play it safe and kick a one-point conversion because they wager that they will score again, kick another one-point conversion, and win in overtime. Suppose the team has the ball with three possessions left and is down fourteen points. If the team follows the one-point conversion strategy, they will win the game only by scoring a touchdown on two successive possessions, holding their opponent scoreless on a single possession and winning in overtime. The probability of winning in this fashion is $(.19)^2(.68)(.5) = .012274$.

If the team goes for two points (as table 24.1 indicates), after scoring with three possessions left they can win the game in one of two ways:

- The first two-point conversion try succeeds, and the team holds the opponent scoreless, scores a touchdown, and makes a one-point conversion.
- The first two-point conversion fails, and the team holds its opponent scoreless, scores a touchdown, makes a two-point conversion, and wins in overtime.

The sum of these two probabilities of winning is

$$0.19 \times 0.19 \times 0.42 \times 0.68 + 0.19 \times 0.19$$
$$\times 0.42 \times 0.58 \times 0.68 \times 0.5 = .0133 > .012274.$$

The probability of multiple independent events occurring is simply the product of the probabilities of each individual event (see chapter 16). Thus, teams should go for two points when down by fourteen points and there is little time left in the game.

Suppose there are at least five possessions remaining and our team is down by four points and scores a touchdown. Most coaches reason that going for three points puts us ahead by one point, so a field goal will not beat us. If there is little time left in the game this logic is correct. If enough time remains, however, the following scenario has a reasonable chance of occurring: the opposition scores a touchdown and a one-point conversion. Then our team is down by four points and scoring a field goal will not allow us to tie the game. If our team is down four points and goes for two points, however, a successful two-point conversion ensures that if the opposition scores a touchdown and a one-point conversion then our team can still tie the game with a field goal.

TO GIVE UP THE BALL IS BETTER
THAN TO RECEIVE

The Case of College Football Overtime

In college football tie games are resolved with overtime. The winner of a coin toss chooses whether to start with the ball or to give the ball to his opponent. The first team with the ball begins on the opponent's 25-yard line and keeps going until they attempt a field goal, score a touchdown, or lose possession. Then the other team gets the ball on their opponent's 25-yard line and keeps going until they attempt a field goal, score a touchdown, or lose possession. The team that is ahead at this point is the winner. If the score is still tied, the order of possessions is reversed and the process is repeated. After the second overtime each team must attempt a two-point conversion. Rosen and Wilson tabulated the outcomes for all overtime games through 2006 and found that the team that had the ball second won 54.9% of the time.[1] This would indicate that the coach who wins the toss should elect to give the other team the ball. The intuitive appeal of giving the ball up is that when you finally get the ball you will know what you need to do to win or keep the game going. For example, if the team that has the ball first scores a touchdown, the other team must go for a touchdown. If the team with the ball first fails to score, then the other team needs a field goal to win. Can we model the "flexibility" of the team that possesses the ball second in a way that is consistent with Rosen and Wilson's findings?

Rosen and Wilson give the following parameter values:

- probability team with ball first scores touchdown: .466
- probability team with ball first scores field goal: .299
- probability team with ball first does not score: .235

[1] Rosen and Wilson, "An Analysis of the Defense First Strategy in College Football Overtime Games."

We will use the term "first team" to refer to the team that gets the ball first and the term "second team" for the team that gets the ball second. We will model the flexibility of the second team by estimating the following two parameters.

- EXTRAFG = Given that the first team scores a field goal, the fraction of possessions during which the second team would convert a possession that would have resulted in no score into a field goal. The rationale here is that if the first team does not score then the second team wins with a field goal and does not need a touchdown. Therefore, they will play more conservatively and be less likely to commit a turnover or be sacked. This will convert some possessions that would have resulted in no score into field goals.
- PRESSURE TD = Probability that the second team will score a touchdown given that the first team has scored a touchdown.

The rationale here is that if the first team scores a touchdown, then the second team will never go for a field goal and some of the possessions that would have resulted in field goals will now result in touchdowns.

Note that if the game is tied after each team has a possession, then the first team has the ball second on the next possession and now has a .549 chance of winning. The first team's chance of winning the game may now be calculated by summing the following probabilities:

- No team scores on their first possession and the first team wins in a later sequence. The probability that the first team does not score on their first possession is .235. The second team knows they only need a field goal, so a fraction EXTRAFG of possessions that would have resulted in no score will now result in a score. At the start of the second possession sequence, the first team is now the second team so they have a 0.549 chance of winning. Therefore, the first team wins in this way with probability $(.235)(.235)(1 - \text{EXTRAFG})(.549)$.
- Each team scores a field goal on their first possession and the first team wins on a later possession. The probability that each team makes a field goal on the first sequence is .299, so the first team wins in this way with probability $(.299)^2(.549) = .0491$.
- The first team scores a field goal and the second team does not score on their first possession. The first team scores a field goal on their first possession with probability .299. The second team fails to score with probability .235. Therefore, the probability that the first team wins in this fashion is $(.299)(.235) = .070$.

- The first team scores a touchdown on their first possession and second team does not score on their first possession. The probability the first team scores a touchdown on their first possession is .469. Since the second team knows they need to score a touchdown their chance of scoring a touchdown will increase because they will never try a field goal (we called the probability of touchdown for the second team PRESSURE TD). Therefore, the probability of the first team winning in this way is $.469 \times (1 - \text{PRESSURE TD})$.

- Both teams score a touchdown on their first possession and the first team wins on a later possession. The first team scores on their first possession with probability .469 and the second team scores on their first possession with probability PRESSURE TD. Then the first team has a .549 chance of winning. Therefore, the chance of the first team winning in this way is .469(PRESSURE TD)(.549).

If the second team can convert 30% of no score possessions into field goals when they know they need only a field goal to win and can score a touchdown with probability 74% (instead of 46.9%) when they know they need a touchdown, then they will win 54.9% of their games. This is fairly consistent with the observed frequency with which the second team wins the overtime.

College football overtime coin toss "strategy" shows the importance of "managerial flexibility." Practitioners of real option theory in finance have long realized that options such as expansion, abandonment, contraction, and postponement of a project have real value.[2] Here we see that the option to go for a field goal instead of a touchdown (or vice versa) can have real value in college football.

[2] Richard Shockley, *An Applied Course in Real Options Valuation* (South-Western Publishing, 2007).

26

WHY IS THE NFL'S OVERTIME SYSTEM
FATALLY FLAWED?

When an NFL game goes into overtime a coin toss takes place and the team that wins the coin toss has the choice of kicking off or receiving. Since the overtime is sudden death, the team winning the coin toss invariably chooses to receive so they have the first chance to score and win the game. During the 1994–2006 seasons the team that received the kickoff in overtime won 60% of the games. It seems unfair that in NFL overtime the team winning the coin flip should have such a huge edge. In attempt to lessen the impact of the coin flip result on the game's outcome, the NFL recently proposed moving the kickoff from the 30- to the 35-yard line. This would give the team receiving the kickoff slightly worse field position and theoretically would decrease the chance that the team receiving the kickoff would score on the first possession. This should give the team kicking off a better chance to win the game. As we will see, a simple mathematical analysis (our model is a simplified version of Jones's model)[1] indicates that it will be difficult to give each team an equal chance to win in sudden death if the overtime begins with a kickoff.

A Simple Mathematical Model
of Sudden Death Overtime

Let p be the probability that an average NFL team scores on a possession. Assuming that each team has a probability p of scoring on each possession, what is the probability that the team receiving the kickoff will win the game? During the regular NFL season, overtime games last only one quarter. If no team scores during the first overtime session, the game is a tie. This happens

---

[1] M. A. Jones, "Win, Lose, or Draw: A Markov Chain Analysis of Overtime in the National Football League," *College Mathematics Journal* 35 (November 2004): 330–36.

less than 5% of the time. Our analysis is greatly simplified if we assume that the overtime can theoretically continue forever. Since less than 5% of NFL games fail to yield a winner during the overtime, our assumption differs from reality in, at most, 5% of all games, so our assumption should not cause our calculations to differ too much from reality. Let K = probability that the team receiving the kickoff in overtime wins the game. There are two ways the receiving team can win:

- With probability p the receiving team scores on the first possession.
- The receiving team fails to score on the first possession, the kicking team fails to score on their possession, and the receiving team wins on a later possession. Assuming the outcomes of the first two possessions and later possessions are independent, the probability that the receiving team wins in this fashion is $(1 - p)(1 - p) \times K$. This follows because if overtime can go on forever, the receiving team has the same chance of winning at the beginning of their second possession as they do at the beginning of their first possession.

Therefore we find that $K = p + (1 - p)(1 - p)K$. Solving for K we find that

$$K = \frac{p}{1-(1-p)^2} = \frac{1}{2-p}. \tag{1}$$

The first thing to note is that since $p < 1$, K must be greater than 0.5. Therefore, if our simple model approximates reality, there is no way for the NFL to make a sudden death format beginning with a kickoff fair.

Does Our Model Approximate Reality?

Our model assumes a (possibly) infinite overtime period and also ignores the possibility that the game will end on a turnover touchdown, safety, or kick or punt return touchdown. Despite this fact, our model predicts (correctly) that the receiving team will win 60% of the overtime games. We can estimate p using NFL data. During the 2003–6 seasons each team averaged around 6,000 possessions. On average, 1,108 of these possessions resulted in rushing or receiving touchdowns and 755 resulted in field goals. Therefore we estimate

$$p = \frac{1,108 + 755}{6,000} = .31.$$

Equation (1) now implies that the probability that the receiving team wins in overtime is given by $\dfrac{1}{2-.31}=.59$. Since receiving teams have triumphed in 60% of all overtimes, our model appears quite consistent with reality.

Of course, if the NFL moved the kickoff position a lot, then our assumption that each team has the same chance of scoring on each possession would be inaccurate. We will explore the idea of changing the kickoff position later in the chapter.

Is There a Fair Solution to the Overtime Dilemma?

We have seen that both the NCAA and NFL overtime approaches give an edge to the team winning the coin toss. Are there some reasonable solutions that could give each team an equal shot at winning in overtime? Professors Jonathan Berk and Terry Hendershott of the Haas Business School at Berkeley suggest that each team "bid" for the yard line on which their first possession starts.[2] The team that "bids" closer to their goal line wins. For example, if the Colts bid for the 20-yard line and the Patriots bid for the 30-yard line, then the Colts begin the overtime with the ball on their 20-yard line. If they fail to score and later lose the game, they have only themselves to blame. This approach allows teams to bid based on their strengths and weaknesses. A good offensive team like the Colts is probably confident they can drive to field goal range and will bid close to their own goal line to ensure possession. A good defensive team will probably be happy to let the other team start with ball deep in their own territory. If there is a tie on the first bid then the teams each submit another bid.

Another fair solution to the NFL overtime problem is analogous to the solution to the famous cake-cutting problem.[3] Consider two people who want to cut a cake "fairly" into two pieces. The mathematically fair solution is to have one person cut the cake into two pieces and have the other person choose which piece he wants to eat. The cake cutter's solution to NFL overtime would be to toss the coin and let the winner choose a yard line on which to begin the first possession or let his opponent choose a yard line on which to begin the first possession. The team that does not choose the yard line gets to

[2] Bialik, "Should the Outcome of a Coin Flip Mean So Much in NFL Overtime?"

[3] See Steven J. Brams, *Mathematics and Democracy: Designing Better Voting and Fair-Division Procedures* (Princeton University Press, 2007), for a wonderful discussion of cake-cutting problems, voting methods, and other interesting social and political issues.

choose whether or not to take the ball. This forces the team that chooses the yard line to choose a "fair" starting situation. For example, if I choose the 50-yard line, my opponent will surely choose the ball. If I choose my 10-yard line, the opponent will probably let me start from my 10-yard line. Assume that I am allowed to choose the yard line on which the first possession starts. Assuming the two teams are of equal ability, I would pick the yard line x for which I feel I have a 50% chance of winning the game if I start with the ball on yard line x. If the other side gives me the ball, I have a 50% chance of winning. Assume the teams are equal. Then the other team also has a 50% chance of winning if they have the ball first and start on yard line x. Thus my choice guarantees me a 50% chance of victory. If my opponent chooses yard line x, then I choose to start with the ball if I believe my chance of winning from yard line x is at least .5, and I choose to give the other team the ball if I believe that my chance of winning from yard line x is less than 0.5. Assuming the two teams are of equal ability, then this choice again ensures that I have at least a 50% chance of winning. Therefore, this solution ensures that whether or not a team wins a coin flip, they have a 50% chance of winning. Of course, our argument shows that if we win the coin flip, then we should let the opponent choose the yard line. Then our chance of winning might exceed 50%.

How About Moving the Kickoff?

Both of these nontraditional solutions are fair and based on sound economic and mathematical theory. Let's hope the NFL sees the light and adds fairness to the excitement that accompanies any sudden death overtime.

For those of you who are old school and demand that an NFL overtime begin with a kickoff, David Romer suggests moving the kickoff from the 30-yard line to a point closer to the opponent's goal line.[4] This would give the receiving team worse field position and would lessen the receiving team's chance of winning the game. He believes that moving the kickoff to the 35-yard line would reduce the receiving team's edge from 60% to 55%. If we had data on all NFL possessions we could calculate the kickoff position that equalizes each team's chance of winning in the following way. For each possession starting on yard line x, determine the fraction of the time that the team with the ball scores first. Find the x for which this probability is .5. Burke has shown that if the kickoff team kicks off from y yards from

---

[4] Romer, "It's Fourth Down and What Does the Bellman Equation Say?"

their goal line, then on average the receiving team receives the ball $42.4 - .5y$ yards from their goal line.[5] For example, if we find $x = 20$, then choosing $y = 45$ would make the receiving team on average start from the 22-yard line and make the game fair.

[5] See http://www.bbnflstats.com/2007/12/best-defensive-player-in-nfl-is-neil.html.

27

HOW VALUABLE ARE HIGH DRAFT PICKS IN THE NFL?

The NFL is generally thought to exhibit more parity than other leagues. This means that it appears easier for a bad NFL team to improve from season to season than for a bad NBA or MLB team to improve from one season to the next. We will investigate the truth of this matter in chapter 41.

Most NFL fans believe that the major equalizer from year to year is the structure of the NFL draft. Teams draft in each round in inverse order of performance with the worst team getting the first pick and the best team getting the last pick. Common sense tells us that an earlier draft pick should, on average, be a more valuable player to a team than a later pick. According to Thaler and Massey (TM), common sense may be wrong.[1]

Estimating the NFL Implied Draft Position Value Curve

TM began by trying to estimate the relative value NFL teams associate with different picks. They collected data on all draft day trades from recent years in which draft picks were dealt. For example, perhaps one team traded their ninth and twenty-fifth picks for a third pick. Letting $v(n)$ be the relative value ($v(1) = 1$) of the nth pick in the draft, this trade would indicate that NFL teams believe $v(9) + v(25) = v(3)$. TM found that the function

$$v(n) = e^{-a(n-1)^b} \tag{1}$$

yields an excellent fit with the observed trade data. For each trade they calculated the estimated value according to (1) of both sides of the trade. Then

---

[1] Thaler and Massey, "The Loser's Curse."

they chose the two constants (a and b) in (1) to minimize the sum (overall observed trades) of the squared difference between the estimated values of both sides of the trade. Thus for a trade in which pick 3 was traded for picks 9 and 25, we would try to choose a and b to minimize $(v(3) - (v(9) + v(25)))^2$. They found that the constants a = $-.148$ and b = $.7$ caused (1) (an instance of the Weibull function) to minimize the sum of the squared differences over all trades. This function is shown in figure 27.1.

TM then tried to determine how much a draft pick benefits a team. They divided a player's performance during a season into five categories:

- not on roster
- no starts
- 1–8 starts
- 9–16 starts
- Pro Bowler

TM then found the average salary by position for each of the five categories. For example, table 27.1 illustrates the performance and value correlations for the quarterback position.

For each draft pick we can compute a player's surplus as player value−player salary. For example, a Pro Bowl quarterback who was paid $5 million earned a surplus of $4,208,248. TM found that the average surplus by position in the draft increased through pick 43, which means that later picks contributed more value than earlier picks. TM view the fact that later

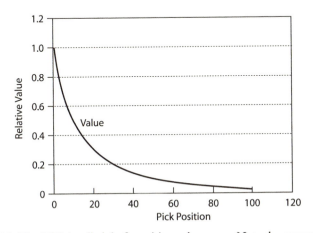

Figure 27.1. The NFL implied draft position value curve. Note the steepness of the curve. For example, pick 10 is only considered half as valuable as pick 1.

TABLE 27.1
Mean Salary and Performance Level
for Quarterback

Mean Salary	Performance Level
$ —	not on roster
$ 1,039,870	no starts
$ 1,129,260	1 to 8 starts
$ 4,525,227	> 8 starts
$ 9,208,248	Pro Bowl

picks contribute more surplus value to teams as evidence of market inefficiency. This conclusion would seem to indicate that NFL teams are not
very proficient at selecting college players.

Phil Birnbaum's Critique of TM's Results

As pointed out by Birnbaum,[2] there is a major flaw in the TM analysis. TM
assume that all players who play the same position and are in the same performance category are equally valuable. For example, TM would assume
that since Joseph Addai started fewer than eight games for the Colts during
2006, his performance was equivalent to any running back who started a
single game. As we saw in chapter 22, Addai had a great year in 2006. Thus,
the TM approach foolishly equates the great Addai with a player who might
have carried the ball twenty times and started one game. In order for TM to
nail down their conclusion that the NFL draft is inefficient, they need a better measure of player performance.

In chapter 33 we will evaluate the efficiency of the NBA draft. We will
develop a much more precise measure of player value than the crude measure used by TM. Our results will show that the NBA draft exhibits much
more efficiency than the high degree of inefficiency found by TM.

---

[2] Phil Birnbaum, "Do NFL Teams Overvalue High Draft Picks?" December 5, 2006,
http://sabermetricresearch.blogspot.com/2006/12/do-nfl-teams-overvalue-high-draft-picks
.html.

Using the NFL Combine 40-Yard Dash to Predict
Running Back Performance

Whether TM are correct about the poor decision-making exhibited by NFL draft selectors, it is clear that teams would benefit if they could more accurately identify good and bad draft picks. Bill Barnwell has recently shown how a prospective draftee's 40-yard dash time can be a good predictor of a running back's NFL success.[3] Barnwell found that a running back's 40-yard dash time in the NFL combine had a $-.36$ correlation with the back's yards gained and carries. This negative correlation indicates (surprisingly) that faster runners tend to perform more poorly than slower runners. Then Barnwell astutely realized that basic physics would indicate that a slightly slower heavy runner might be more effective than a slightly faster lighter runner. After analyzing the 40-yard dash times and NFL performances of all running backs who participated in the NFL combine during the time frame 1999–2008, Barnwell found that

$$\text{weight-adjusted 40-yard dash score} = \frac{200 \times \text{weight in pounds}}{(40 \text{ yard time in seconds})^4}$$

has a .45 correlation with yards and carries. Of course the WA40 (short for weight-adjusted 40-yard dash score) increases as a player gets heavier and faster. WA40 is calibrated so that it averages out to one hundred for all NFL running backs.

During 1999–2008, thirty running backs were chosen in the first round. Twenty-eight of them had WA40 > 1. The only two first-round running back draftees who had WA40 < 1 played poorly in the NFL (Trung Canidate and William Green).

The best WA40 during the 1999–2008 time frame was that of the Giants' current starting running back, Brandon Jacobs. Jacobs weighed 267 pounds and ran a 4.56 40-yard dash. Therefore, his WA40 score is $\frac{200(267)}{(4.56)^4} = 122.42$. Of course, Jacobs was a key performer in the Giants' 2007–8 drive to the Super Bowl.

---

[3] Smith, "The Race of Truth."

Barnwell found that a running back's vertical jump had a 0.28 correlation with performance, while performance in the bench press, broad jump, shuttle drill, and three cone drill were virtually uncorrelated to future NFL performance.

I am sure that data miners will develop many ways to use available data (NFL combine performance and college statistics) to increase the accuracy of NFL draft selectors.

The Winner's Curse

If we accept that the NFL draft is inefficient, what might cause the observed inefficiency? Perhaps this is an instance of the well-known Winner's Curse.[4] Essentially the Winner's Curse says that winners of auctions often pay more than the object they won is worth. The Winner's Curse was first observed during bidding for offshore oil leases during the 1950s. Many companies that won the rights to offshore oil sites found that the value of the site was on average less than the value they bid for the site. Suppose the true value of a prize is $100 million (of course, nobody knows this for sure). If five people bid for the prize they might value the prize at $60, $80, $100, $120, and $140 million. (Note these values average to the true value of $100 million, as we would expect.) The person placing the $140 million value on the prize will probably bid near $140 million for the prize. The "high-value person" will win the prize but lose around $40 million. In the NFL the team that drafts a player has, in a sense, "bid the highest" for that player, so if the Winner's Curse is at play here, it is reasonable to expect that earlier draft picks may create less value than do later draft picks.

---

[4] Richard Thaler, *The Winner's Curse* (Princeton: Princeton University Press, 1997).

PART III

BASKETBALL

BASKETBALL STATISTICS 101

The Four-Factor Model

For each player and team NBA box scores track the following information:

- two-point field goals made and missed
- three-point field goals made and missed
- free throws made and missed
- personal fouls committed
- assists
- offensive and defensive rebounds
- blocked shots
- turnovers
- steals
- minutes played

How can we use this data (on a per game or per season basis) to break down what makes an NBA team perform well or poorly?

Effective Field Goal Percentage

Many coaches and players currently evaluate their shooting by looking at their field goal percentage. For example, suppose in a Dallas Mavericks–New York Knicks game the Mavericks make 45 out of 100 field goals (shoot 45%). The Knicks make 50 out of 100 field goals (shoot 50%). At first glance the Knicks shot better than the Mavericks. Suppose, however, the Mavericks shot 15 for 20 on three-pointers and the Knicks shot 1 for 5 on three-pointers. On the same number of shots the Mavericks scored 105 points and the Knicks scored 101 points. This indicates that the Mavericks actually shot better from the field than the Knicks did. To capture this phenomenon, NBA statistical geeks have created Effective Field Goal percentage (EFG).

EFG = (all field goals made + 0.5 (3-point field goals made)/
          (all field goal attempts).

In essence, EFG gives 50% more credit for making a three-pointer because a three-pointer is worth 50% more points than a two-point field goal. For our example, the Mavericks' EFG = $(45 + 0.5 \times (15))/100 = 52.5\%$ and the Knicks' EFG = $(50 + 0.5 \times (1))/100 = 50.5\%$. This example shows that EFG does a better job of capturing the quality of a team's shooting than does the traditional metric of Field Goal Percentage.

How Can We Evaluate Team Rebounding?

Raw rebounds for a team or player can be misleading. What really matters is the percentage of rebounds a team gets when they are on offense (called Offensive Rebounding Percentage, or ORP) and the percentage of rebounds a team gets when they are on defense (called Defensive Rebounding Percentage, or DRP).

Dean Oliver, a statistical consultant for the Denver Nuggets, describes a four-factor model that can be used to analyze a team's performance and to better understand a team's strengths and weaknesses.[1] The four factors (measured for both the team's offense and the team's defense) are explained below.

Four Factors for Team Offense
1. Effective Field Goal Percentage (EFG).
2. Turnovers Committed per Possession (TPP). A possession starts when a team gets the basketball and ends when they give up control of the basketball. During the last five seasons NBA teams have averaged around 92 possessions per game.[2]
3. Offensive Rebounding Percentage (ORP): the percentage of rebounds a team gets of their missed shots.
4. Free Throw Rate (FTR): foul shots made divided by field goal attempts. FTR is impacted by how often a team gets to the foul line as well as by their free throw percentage.

---

[1] Oliver, *Basketball on Paper*.
[2] See Kubatko et al., "A Starting Point for Analyzing Basketball Statistics."

Four Factors for Team Defense

1. Opponent's Effective Field Goal Percentage (OEFG).
2. Defensive Turnovers Caused per Possession (DTPP).
3. Defensive Rebounding Percentage (DRP): percentage of rebounds a team gets of their opponent's missed shots.
4. Opponent's Free Throw Rate (OFTR): foul shots made by the opposing team divided by field goal attempts made by the opposing team.

Figure 28.1 shows the values of the four factors for teams during the 2006–7 NBA season. Figure 28.2 shows each team's ranking for each of the four factors as well as the number of games won by each team.

	J	K	L	M	N	O	P	Q	R	S
3		Mean	0.496166667	0.496166667	0.159	0.158866667	0.270866667	0.729066667	0.246366667	0.246633333
4	Team	Wins	Offensive shooting	Defensive shooting	Offensive TOs	Defensive TOs	Offensive rebounding	Defensive rebounding	Offensive FTs	Defensive FTs
5	76ers	35	0.48	0.501	0.163	0.166	0.272	0.708	0.255	0.221
6	Bobcats	33	0.48	0.5	0.155	0.165	0.264	0.716	0.236	0.279
7	Bucks	28	0.504	0.522	0.158	0.164	0.276	0.681	0.209	0.234
8	Bulls	49	0.493	0.473	0.166	0.18	0.286	0.743	0.229	0.252
9	Cavaliers	50	0.484	0.48	0.152	0.163	0.297	0.758	0.223	0.243
10	Celtics	24	0.479	0.502	0.172	0.161	0.27	0.738	0.259	0.28
11	Clippers	40	0.481	0.488	0.162	0.148	0.272	0.747	0.28	0.249
12	Grizzlies	22	0.504	0.529	0.168	0.157	0.259	0.711	0.285	0.237
13	Hawks	30	0.471	0.503	0.17	0.163	0.292	0.709	0.263	0.268
14	Heat	44	0.506	0.485	0.159	0.154	0.249	0.733	0.222	0.232
15	Hornets	39	0.479	0.499	0.154	0.146	0.291	0.747	0.215	0.212
16	Jazz	51	0.502	0.496	0.164	0.159	0.317	0.751	0.283	0.314
17	Kings	32.4	0.491	0.513	0.149	0.166	0.231	0.725	0.289	0.24
18	Knicks	33	0.494	0.504	0.181	0.145	0.31	0.74	0.27	0.246
19	Lakers	42	0.511	0.5	0.159	0.151	0.261	0.723	0.249	0.262
20	Magic	40	0.5	0.48	0.183	0.162	0.293	0.737	0.276	0.286
21	Mavericks	67	0.509	0.477	0.151	0.157	0.287	0.75	0.256	0.265
22	Nets	41	0.504	0.49	0.157	0.152	0.246	0.744	0.245	0.266
23	Nuggets	45	0.501	0.499	0.164	0.162	0.289	0.718	0.268	0.203
24	Pacers	35	0.474	0.491	0.17	0.165	0.284	0.727	0.246	0.271
25	Pistons	53	0.488	0.477	0.135	0.162	0.283	0.709	0.237	0.234
26	Raptors	47	0.504	0.503	0.142	0.159	0.222	0.745	0.239	0.219
27	Rockets	52	0.499	0.466	0.152	0.15	0.257	0.77	0.22	0.23
28	Sonics	31	0.499	0.515	0.162	0.158	0.278	0.709	0.228	0.243
29	Spurs	58	0.521	0.471	0.15	0.155	0.242	0.757	0.235	0.201
30	Suns	61	0.551	0.492	0.147	0.152	0.227	0.719	0.215	0.206
31	Trail-blazers	32	0.483	0.508	0.163	0.15	0.282	0.73	0.241	0.267
32	Warriors	42	0.512	0.506	0.157	0.182	0.256	0.696	0.215	0.264
33	Wizards	41	0.491	0.517	0.141	0.162	0.281	0.71	0.272	0.249
34	Timber-wolves	32.4	0.49	0.498	0.164	0.15	0.252	0.721	0.231	0.226

Figure 28.1. Four-factor analysis of the NBA 2006–7 season. Source: http://www.basketball-reference.com.

	J	K	T	U	V	W	X	Y	Z	AA
3										
4	Team	Wins	Offensive shooting	Defensive shooting	Offensive TOs	Defensive TOs	Offensive rebounding	Defensive rebounding	Offensive FTs	Defensive FTs
5	76ers	35	25	19	19	3	16	28	12	6
6	Bobcats	33	25	17	11	5	19	22	19	27
7	Bucks	28	7	29	14	7	15	30	30	10
8	Bulls	49	17	3	24	2	9	10	22	19
9	Cavaliers	50	22	6	8	8	3	2	24	14
10	Celtics	24	27	20	28	14	18	12	10	28
11	Clippers	40	24	9	17	28	16	6	4	17
12	Grizzlies	22	7	30	25	18	21	23	2	12
13	Hawks	30	30	21	26	8	5	25	9	25
14	Heat	44	6	8	15	21	25	14	25	9
15	Hornets	39	27	15	10	29	6	6	27	4
16	Jazz	51	11	13	21	15	1	4	3	30
17	Kings	32.4	18	26	5	3	28	17	1	13
18	Knicks	33	16	23	29	30	2	11	7	16
19	Lakers	42	4	17	15	24	20	18	13	20
20	Magic	40	13	6	30	10	4	13	5	29
21	Mavericks	67	5	4	7	18	8	5	11	22
22	Nets	41	7	10	12	22	26	9	15	23
23	Nuggets	45	12	15	21	10	7	21	8	2
24	Pacers	35	29	11	26	5	10	16	14	26
25	Pistons	53	21	4	1	10	11	25	18	10
26	Raptors	47	7	21	3	15	30	8	17	5
27	Rockets	52	14	1	8	25	22	1	26	8
28	Sonics	31	14	27	17	17	14	25	23	14
29	Spurs	58	2	2	6	20	27	3	20	1
30	Suns	61	1	12	4	22	29	20	27	3
31	Trail-blazers	32	23	25	19	25	12	15	16	24
32	Warriors	42	3	24	12	1	23	29	27	21
33	Wizards	41	18	28	2	10	13	24	6	17
34	Timber-wolves	32.4	20	14	21	25	24	19	21	7

Figure 28.2. NBA team rankings by four factors, 2006–7. Source: http://www .basketball-reference.com.

For example, the Mavericks had an EFG of 50.9% and held their opponents to an EFG of 47.7%. The Mavs committed 0.151 turnovers per possession and caused 0.157 turnovers per possession. The Mavs rebounded 28.7% of their missed shots and 75% of their opponents' shots (or, equivalently, their opponents rebounded 25% of their missed shots). The Mavs made 0.256 free throws per FGA while their opponents made 0.265 free throws per FGA. This shows the Mavs bested their opponents on all factors except free throws. This is not surprising because the Mavs are primarily a jump shooting team that does not often drive to the basket.

The Four Factors Are Virtually Uncorrelated

The interesting thing about the four factors is that there is little correlation among them (See chapter 5 for an explanation of correlation). Recall that correlations are always between −1 and +1 with a correlation near +1 for two quantities x and y indicating that when x is big y tends to be big and a correlation near −1 indicating that when x is big y tends to be small. Figure 28.3 shows the correlations found using the Correlation option from Excel's Data Analysis Toolpak.

Notice that most of the correlations are near 0. For example, the correlation between offensive shooting and defensive turnovers is −.10. This indicates that if a team is better than average on offensive shooting, they will be slightly worse than average in causing defensive turnovers. Let's examine the three largest (in absolute value) correlations shown in figure 28.3.

- There is a −.67 correlation between defensive shooting percentage and defensive rebounding. This means that teams that give up a high shooting percentage tend to be poor defensive rebounding teams. This is reasonable because if a team fails to rebound its opponents' missed shots, they will in all likelihood get many easy inside shots or dunks on follow-up shots.
- There is a −.47 correlation between offensive shooting and offensive rebounding. This means that good shooting teams tend to be poor offensive

	A	B	C	D	E	F	G	H	I
1		Offensive shooting	Defensive shooting	Offensive TOs	Defensive TOs	Offensive rebounding	Defensive rebounding	Offensive FTs	Defensive FTs
2	Offensive shooting	1	−0.10823119	−0.27223683	−0.1032321	−0.471982212	−0.000659146	−0.242918	−0.31165359
3	Defensive shooting	−0.10823119	1	0.12781981	0.05038423	−0.047695316	−0.673817819	0.2461414	0.04011147
4	Offensive TOs	−0.272236826	0.127819806	1	−0.0240731	0.455388188	0.003098699	0.3435858	0.41230448
5	Defensive TOs	−0.103232065	0.050384229	−0.02407314	1	0.048171018	−0.397712659	−0.05845	0.22520563
6	Offensive rebounding	−0.471982212	−0.047695316	0.45538819	0.04817102	1	0.057779622	0.2528353	0.44674552
7	Defensive rebounding	−0.000659146	−0.673817819	0.0030987	−0.3977127	0.057779622	1	0.0662925	0.052547
8	Offensive FTs	−0.24291806	0.246141366	0.34358582	−0.0584503	0.252835284	0.066292546	1	0.36660448
9	Defensive FTs	−0.311653593	0.040111475	0.41230448	0.22520563	0.446745517	0.052547005	0.3666045	1

Figure 28.3. Correlations among the four factors.

rebounding teams. The Phoenix Suns (first in offensive shooting and twenty-ninth in offensive rebounding) are an illustration of this phenomenon. Perhaps teams loaded with good shooters do not have the tough guys needed to pound the offensive boards.

- There is a .46 correlation between offensive rebounding and offensive turnovers. This means teams that are good at offensive rebounding also tend to turn the ball over a lot. This is reasonable because good offensive rebounders tend to be poor ball handlers.

Different Paths to Team Success (or Failure)

We can use figure 28.2 to quickly zero in on the keys to success (or failure) for an NBA team. Consider the following examples:

- The 2007 champion San Antonio Spurs won because of their great shooting, holding their opponents to poor shooting opportunities, allowing few turnovers, and by holding down fouls and giving up few points from the foul line.
- The 2007 Phoenix Suns were successful because of great shooting, lack of turnovers, and giving up few points from the foul line. The Suns succeeded despite their poor offensive and defensive rebounding and lack of free throws.
- The 2007 Memphis Grizzlies had the league's worst record. Their shooting percentage defense was poor, they allowed lots of turnovers, and they rebounded poorly.
- The 2007 New York Knicks had a 33–49 record, due almost exclusively to lots of turnovers committed on offense, while their defense caused very few turnovers.

How Important Are the Four Factors?

Can we estimate the relative importance of the four factors? Recall that in chapter 18 we used regression to predict NFL team performance based on measures of passing and rushing efficiency, and of turnover frequency. (See chapter 3 for a description of regression analysis.) We can use regression in a similar way to evaluate the importance of the four factors in basketball performance. Running a regression using the data in figure 28.1 can predict a team's number of wins from the following four independent variables:

- EFG – OEFG
- TPP – DTPP

- ORP – DRP
- FTR – OFTR.

The results of the regression are shown in figure 28.4. From these results we can predict

$$\text{games won} = 41.06 + 351.88(\text{EFG} - \text{OEFG})$$
$$+ 333.06(\text{TPP} - \text{DTPP}) + 130.61(\text{ORP} - \text{DRP})$$
$$+ 44.43(\text{FTR} - \text{OFTR}).$$

	A	B	C	D	E
1	SUMMARY OUTPUT				
3	Regression Statistics				
4	Multiple R	0.953165405			
5	R Square	0.908524289			
6	Adjusted R Square	0.893888175			
7	Standard Error	3.533831698			
8	Observations	30			
9					
10	ANOVA				
11		df	SS	MS	F
12	Regression	4	3100.719505	775.17988	62.0741478
13	Residual	25	312.1991617	12.487966	
14	Total	29	3412.918667		
16		Coefficients	Standard Error	t Stat	P-value
17	Intercept.	41.05829628	0.645276505	63.628996	3.54259E-29
18	Shooting Dev.	351.8800481	28.36841836	12.403936	3.52088E-12
19	Turnover Dev.	333.0598616	52.17592394	6.3834013	1.10625E-06
20	Rebound Dev.	130.6051555	22.96495865	5.6871496	6.37464E-06
21	Free Throw Dev.	44.42983641	23.55911647	1.8858872	0.070980928

Figure 28.4. Four-factor regression.

These four independent variables explain 91% of the variation in the number of games won. The standard error of 3.53 means we are 95% sure our predicted wins will be within $2(3.53) = 7.06$ wins of the actual number of wins.

To measure the impact of the four factors on wins, we can look at the correlations between the four factors and wins:

- EFG – OEFG has a .85 correlation with wins and by itself explains 71% of the variation in wins.
- TPP – DTPP has a 0.38 correlation with wins and by itself explains 15% of variation in wins.
- ORP – DRP has a 0.25 correlation with wins and by itself explains 6% of variation in wins.

- FTR − OFTR has a −0.01 correlation with wins and by itself explains virtually none of the variation in wins.

This analysis indicates that NBA teams' differential on shooting percentage is by far the most important factor in their success.

The relative importance of the four factors are summarized as follows:

- A 0.01 improvement in EFG − OEFG is worth 3.5 wins. That is any of the following improvements:
  1. improve our EFG 1% (say, from 47% to 48%);
  2. reduce our opponent's EFG by 1%; or
  3. improve our EFG by 0.5% and cut our opponent's EFG by 0.5% should on average cause us to win 3.5 more games.

- A 0.01 improvement in TPP − DPPP is worth 3.3 wins. Thus any of the following improvements:
  1. one less turnover per 100 possessions; or
  2. one less turnover per 200 possessions and causing one more turnover per 200 possessions would lead to 3.3 more wins.

- An increase of 0.01 in ORP − DRP would lead on average to 1.3 more wins per season. Thus any of the following combinations would be expected to lead to 1.3 more wins:
  1. one more offensive rebound per 100 missed shots;
  2. one more defensive rebound per 100 shots missed by opponent; or
  3. one more offensive rebound per 200 missed shots and one more offensive rebound per 200 shots missed by an opponent.

- An increase of 0.01 in FTR − OFTR would be expected to lead to 0.44 wins. Therefore any of these three combinations would be expected to lead to 0.44 wins:
  1. one more free throw made per 100 field goal attempts;
  2. one less free throw given up per 100 field goal attempts by opponent; or
  3. one more free throw made per 200 field goal attempts and one less free throw given up per 200 field goal attempts.

In summary, Dean Oliver's decomposition of a team's ability into four factors provides a quick and effective way to diagnose a team's strengths and weaknesses. Of course, the four-factor model may be applied to final or season-in-progress cumulative team data, or simply to the box score data for any particular game.

29

LINEAR WEIGHTS FOR EVALUATING NBA PLAYERS

In chapter 3 we discussed the use of Linear Weights to evaluate MLB hitters. We found that by determining appropriate weights for singles, walks, doubles, triples, home runs, outs, stolen bases, and caught stealing we can do a pretty good job of estimating the Runs Created by a hitter.

Given the wealth of information in an NBA box score, many people have tried to come up with Linear Weights formulas that multiply each box score statistic by a weight and equate the weighted sum of player statistics as a measure of the player's ability. In this chapter we will discuss several linear weighting schemes used to rate NBA players:

- The NBA Efficiency metric
- John Hollinger's PER and Game Score ratings
- Berri, Schmidt, and Brook's (BSB) Win Scores.[1]

NBA Efficiency Rating

Let's begin by discussing the NBA Efficiency rating (created by Dave Heeren), which is computed as

efficiency per game = (points per game) + (rebounds per game)
+ (assists per game) + (steals per game)
− (turnovers per game) − (missed FG per game)
− (missed FT per game).

This simplistic system essentially says that all good statistics are worth +1 and all bad statistics are worth −1. This does not make sense. For example, a player who shoots 26.67% on three-point field goals (which is almost

---

[1] Berri, Schmidt, and Brook, *Wages of Wins*.

10% below the league average) would raise his efficiency by taking more three-pointers. For example, suppose a player shot 5 for 18 on three-pointers. He scored 15 points and missed 13 shots. This player's three-point field goal attempts would yield $15 - 13 = 2$ Efficiency points. If he shot 10 for 36, his three-point field goal attempts would yield $30 - 26 = 4$ Efficiency points. Any player who shot this badly would be told not to shoot! Similarly, consider a player who shoots 36.4% (4 for 11) on two-pointers. If he takes 11 shots, he scores 8 points and misses 7 shots. These shots add $8 - 7 = 1$ point to his Efficiency rating. If he shot 8 for 22 (which is almost 10% below league average) his shots would add 2 points to his Efficiency rating.[2] The 2006–7 leaders in Efficiency rating are listed in table 29.1. In chapter 30 we will see that certain players on this list (such as Carmelo Anthony and Amare Stoudemire) are vastly overrated according to the NBA Efficiency statistic. Other players not on the list (such as Luol Deng and Anthony Parker) are vastly underrated by the NBA Efficiency metric.

The Player Efficiency Rating System

Now let's turn our attention to John Hollinger's well-known PER (Player Efficiency) and Game Score ratings.[3] An average NBA player has a PER score of 15. Hollinger's PER rating formula is complex, but as pointed out by Berri,[4] a player who shoots more than 30.4% on two-point field goals can increase his PER rating by taking more shots. In addition, a player who shoots more than 21.4% on three-point field goals can increase his PER rating by taking more shots. (No regular NBA player shoots this badly.) Thus Hollinger's PER rating implies that if you are by far the worst shooter in the league, you can help your team by taking more shots. We can also see that Hollinger assigns incorrect weights to shooting statistics by looking at his Game Score formula, which is used to rank player performances during a game:

---

[2] The reader should be able to show that a player who shoots below 25% on three-pointers or below 33.33% on two-pointers would improve his Efficiency rating by taking more shots. For example, if a player makes x three-pointers and misses $(100 - x)$ three-pointers, she would contribute $3x - (100 - x)$ to the Efficiency rating. This is positive as long as $4x - 100 > 0$ or $x > 25$. Thus a player shooting >25% on three-pointers increases his efficiency by taking more three-pointers. This is a disquieting property of the Efficiency rating.

[3] Hollinger's ratings can be found on http://espn.com.

[4] See Berri's comments at http://dberri.wordpress.com/2006/11/17/a-comment-on-the-player-efficiency-rating/.

TABLE 29.1
NBA Efficiency Leaders Ranked on Per Game Basis (2006–7 Season)

	PLAYER NAME, TEAM NAME	GP	MPG	PTS	EFF	RPG	APG	STPG	BLKPG	EFF48M	EFF
1	Kevin Garnett, MIN	76	39.2	22.4	29.2	12.8	4.1	1.2	1.66	35.53	29.17
2	Kobe Bryant, LAL	77	40.7	31.6	27.6	5.7	5.4	1.4	0.47	32.55	27.65
3	Dywane Wade, MIA	51	37.5	27.4	27.1	4.7	7.5	2.1	1.22	34.25	27.06
4	Dirk Nowitzki, DAL	78	36.8	24.6	26.9	8.9	3.4	0.7	0.8	35.71	26.9
5	Ming Yao, HOU	48	32.7	25.0	25.8	9.4	2.0	0.4	1.96	36.63	25.81
6	Carlos Boozer, UTA	74	34.6	20.9	25.8	11.7	3.0	0.9	0.28	35.86	25.81
7	LeBron James, CLE	78	40.9	27.3	25.6	6.7	6.0	1.6	0.7	30.05	25.6
8	Pau Gasol, MEM	59	35.4	20.8	25.5	9.8	3.4	0.5	2.14	33.85	25.49
9	Tim Duncan, SAS	80	33.9	20.0	25.4	10.6	3.4	0.8	2.38	35.76	25.39
10	Chris Bosh, TOR	69	38.5	22.6	25.3	10.7	2.5	0.6	1.3	31.59	25.35
11	Elton Brand, LAC	80	38.4	20.5	25.0	9.3	2.9	1.0	2.24	31.18	24.99
12	Steve Nash, PHX	76	35.1	18.6	24.5	3.5	11.6	0.8	0.08	33.32	24.5
13	Gilbert Arenas, WAS	74	39.0	28.4	24.2	4.6	6.0	1.9	0.18	29.19	24.18
14	Shawn Marion, PHX	80	37.4	17.5	24.1	9.8	1.7	2.0	1.52	30.76	24.11

(*cont.*)

TABLE 32.1 (cont.)

	PLAYER NAME, TEAM NAME	GP	MPG	PTS	EFF	RPG	APG	STPG	BLKPG	EFF48M	EFF
15	Amare Stoudemire, PHX	82	32.7	20.4	23.4	9.6	1.0	1.0	1.34	34.29	23.43
16	Carmelo Anthony, DEN	65	37.3	28.9	23.3	6.0	3.8	1.2	0.34	29.28	23.32
17	Dwight Howard, ORL	82	36.2	17.6	23.1	12.3	1.9	0.9	1.9	30.08	23.11
18	Marcus Camby, DEN	70	33.3	11.2	23.0	11.7	3.2	1.2	3.3	32.66	23.04
19	Vince Carter, NJN	82	38.3	25.2	22.7	6.0	4.8	1.0	0.37	28.62	22.73
20	Zach Randolph, POR	68	35.7	23.6	22.4	10.1	2.2	0.8	0.22	30.15	22

$$\begin{aligned}
\text{game score} = {} & (\text{points} \times 1.0) + (\text{FGM} \times 0.4) + (\text{FGA} \times -0.7) \\
& + (\text{FTA} - \text{FTM}) \times -0.4) + (\text{OREB} \times 0.7) \\
& + (\text{DREB} \times 0.3) + (\text{STL} \times 1.0) + (\text{AST} \times 0.7) \\
& + (\text{BLK} \times 0.7) + (\text{PF} \times -0.4) + (\text{TO} \times -1.0).[5]
\end{aligned}$$

A player shooting over 20.4% (0.7/3.4) on three-pointers will increase his Game Score by taking more shots. A player shooting over 29.2% on two-pointers (0.7/2.4) will also increase his Game Score by taking more shots. (This implies that the worst shooter in the league would help his team by taking more shots.)

Berri found a 0.99 correlation between Hollinger's Game Score ratings and his PER ratings.[6] This shows that PER and Game Score will almost always rank players in the same order. Berri also found a 0.99 correlation between Hollinger's Game Score ratings and NBA Efficiency ratings. This suggests that Hollinger's rankings are simply a minor repackaging of NBA Efficiency.

Win Scores and Wins Produced

Now we turn to BSB's Win Scores and Wins Produced. Although the Win Score formula published on their blog[7] is only an approximation of their more complex formula, BSB state that the following formula closely approximates their more complicated method for ranking players.

$$\begin{aligned}
\text{player win score} = {} & \text{points} + \text{rebounds} + \text{steals} + 0.5(\text{assists}) \\
& + 0.5(\text{blocked shots}) - \text{FG attempts} \\
& - \text{turnovers} - 0.5(\text{FT attempts}) \\
& - 0.5(\text{personal fouls}).
\end{aligned}$$

Unlike the NBA Efficiency metric or Hollinger's PER rating, the Linear Weights in the Win Score metric seem much more sensible. For example, to raise his rating by shooting more, a player needs to shoot over 50% on two-point field goals or over 33.33% on three-point field goals. It also seems reasonable to give equal weight to turnovers and rebounds, because a rebound gets you possession and a turnover gives up possession. Using a complex method BSB convert Win Scores into Wins Produced. The nice thing about Wins Produced is that the sum of Wins Produced by all of a

---

[5] FGM = field goals made; FGA = field goals attempted; FTA = free throws attempted; FTM = free throws made; OREB = offensive rebounds; DREB = defensive rebounds; STL = steals; AST = assists; BLK = blocks; PF = personal fouls; TO = turnovers.

[6] Berri, private communication with the author, July 2007.

[7] See http://dberri.wordpress.com/2006/05/21/simple-models-of-player-performance/.

team's players will be nearly the same as the team's total number of wins. For example, for the 2006–7 Chicago Bulls, BSB found the Wins Produced shown in table 29.2. The Bulls actually won 55 games (including playoff games) during the 2006–7 season. BSB believe (and I agree) that Luol Deng was the Bulls' best player and created 15.56 wins. The total Wins Produced for the Bulls is 52.17, which is close to their actual total of 55 wins. However, the fact that for most teams the Wins Produced for each player on a team nearly equal the team's total wins does not imply that the team's "wins" are accurately partitioned among the team's players.

As an example, Bruce Bowen of the Spurs always fares poorly in terms of Win Score and Wins Produced. Most NBA experts agree that Bowen is a superb defender. BSB attempt to acknowledge the fact that Bowen is a

TABLE 29.2
Wins Produced by Player, Chicago Bulls, 2006–7

Wins Produced	Name
15.5561875	Deng, Luol
8.74225	Hinrich, Kirk
4.478541667	Gordon, Ben
16.049625	Wallace, Ben
3.6920625	Duhon, Chris
−0.560333333	Brown, P.J.
2.150416667	Nocioni, Andres
2.077666667	Thomas, Tyrus
1.024833333	Sefolosha, Thabo
−1.809166667	Allen, Malik
1.305	Griffin, Adrian
−0.34375	Sweetney, Michael
−0.213583333	Khryapa, Viktor
0.020416667	Barrett, Andre
0	Andriuskevicius, Martynas

good defender by introducing a "team adjustment" that accounts for the Spurs' excellent defense. The problem with this approach is that the team adjustment does not reveal how much of the Spurs' defensive effectiveness is a result of their excellent defenders (Tim Duncan and Bowen), in contrast to their average defenders (like Manu Ginobili and Tony Parker). This means that Bowen's defensive ability is not properly accounted for by BSB. Because basketball is half offense and half defense, the shortage of defensive statistics in the box scores causes BSB to shortchange great defenders like Bowen.

For most players (other than some defensive stalwarts) I believe that BSB do an excellent job of weighting the box score statistics that are best for player evaluation. The only problem I have with the BSB ratings is that they are based on the statistics tabulated in a box score, so their Win Score metric primarily captures player attributes that are measured in the box score. I am sure that the activity during at least 80% of any game is not tabulated in a box score. For example, the box score does not give credit for the following:

- taking a charge
- deflecting a pass
- boxing out so my teammate gets the rebound
- the pass before the pass that earns an assist
- helping out on defense when my teammate is beaten by a quick guard
- setting a screen that leads to an open three-pointer.

All these events help the team, but none is reflected in the box score. In the next chapter we discuss the rating of players by Adjusted $+/-$ ratings. This approach has been used by the Dallas Mavericks to rate NBA players since the 2000–2001 season. Adjusted $+/-$ ratings are based on the premise that a good player helps his team. It is not concerned with a player's box score statistical contribution; it simply focuses on how the player's team does when he is on the court, versus how he does when he is off the court.

30

ADJUSTED +/− PLAYER RATINGS

Basketball is a team game. The definition of a good player is somebody who makes his team better, not a player who scores 40 points per game. My favorite story about what defines a good player is told by Terry Pluto in his excellent book *Tall Tales*.[1] The late great Celtics coach Red Auerbach said that whenever the Celtics practiced, KC Jones's team always won. This must mean he was a good or great player. Yet during his peak years his PER rating was around 10, indicating he was a poor player. KC must have done some great things that did not show up in the box score.

Pure +/− Ratings and Their Flaws

The first statistic that tied a player rating to team performance was hockey's +/− statistic. A player's +/− statistic is simply how many goals a player's team outscores their opponents by when the player is on the ice (power plays, when one team has more players on the ice, are excluded from the calculation). Since 1968 the NHL has kept track of +/− for each player. The highest recorded +/− was that of Bobby Orr, who during the 1967–68 season earned a +124 goal +/− .

For NBA players, Pure +/− per 48 minutes on court was thought to be a valid measure of how a player helped his team.[2] The problem with Pure +/− statistics is that a player's Pure +/− statistic depends on the quality of the players he plays with and against. Suppose Player A had a Pure +/− of 0 and played for the 2006–7 Memphis Grizzlies, the league's worst team.

---

[1] Terry Pluto, *Tall Tales: The Glory Years of the NBA* (Bison Books, 2000).

[2] Since 2003 +/− statistics have been available 2003 at http://www.82games.com, and beginning with the 2006–7 season, Pure +/− statistics have been available on http://www.nba.com (http://www.nba.com/statistics/lenovo/lenovo.jsp). As of November 2007, nba.com posts each player's individual +/− updated while the game is being played.

TABLE 30.1
Pure +/−(Includes Playoff Games)

1.	Devin Harris—Dallas:	+11.48
2.	Tim Duncan—San Antonio:	+11.48
3.	Manu Ginobili—San Antonio:	+11.39
4.	Steve Nash—Phoenix:	+10.33
5.	Eric Dampier—Dallas:	+10.20
6.	Bruce Bowen—San Antonio:	+9.58
7.	Dirk Nowitzki—Dallas:	+9.42
8.	Shawn Marion—Phoenix:	+9.34
9.	Tony Parker—San Antonio:	+8.82
10.	Josh Howard—Dallas:	+8.42

Suppose Player B also had a Pure +/− of 0 and played for the 2006–7 NBA champion San Antonio Spurs. Who is the better player? Certainly Player A. Player A made a terrible team average. Player B made a great team average. Table 30.1 illustrates how unadjusted (or Pure) +/− statistics are fairly useless. The table lists the players (who played at least twenty minutes a game) who had the ten best Pure +/− during the 2006–7 season. All played for the top three teams in the league (Spurs, Mavs, and Suns). Even Eric Dampier and Bruce Bowen's agents would not say they are among the ten best players in the league. Dampier and Bowen have great +/− ratings because they are usually on the court with great players like Devin Harris, Dirk Nowitzki, Tim Duncan, or Manu Ginobili. As further proof that Pure +/− is virtually useless, during the 2006–7 season no player on a team with a .500 or worse record was one the top 40 players based on Pure +/− . Gilbert Arenas, for example, ranks #45 based on Pure +/− . Kevin Garnett (more on Garnett's greatness later in the chapter) had a Pure +/− of only + .14.

Adjusted +/−: A Better Approach

The trick to making sense of Pure +/− is to adjust each player's rating based on the ability of the players he is on the court with and the players he plays against. The data used to make this adjustment consist of (by the end of the season) over 38,000 rows of play-by-play data. Each row represents a segment of time during which the players on the court remain unchanged. For example, a sample time segment might look like the following:

Dallas, Home against San Antonio
On court for Dallas: Howard, Nowitzki, Harris, Terry, Dampier
On court for San Antonio: Duncan, Parker, Ginobili, Barry, Oberto
Length of time segment: 3 minutes
Score during time segment: Dallas ahead 9–7

We adjust the score during each time segment based on a home edge of 3.2 points per 48 minutes. Thus our adjusted score would be Dallas ahead $9 - (3/48) \times .5(3.2)$ to $7 + (3/48) \times 0.5(3.2)$, or Dallas ahead 8.9–7.1.

We have a variable for each player that represents his Adjusted $+/-$ rating per 48 minutes. An average NBA player will have an Adjusted $+/-$ of 0. Thus an Adjusted $+/-$ rating of 5 means that if the player replaced an average NBA player for 48 minutes, his team would improve by an average of 5 points per game. For any time segment we predict the home team's margin of victory per minute to be $(3.2/48) +$ (sum of home team player ratings for players on court) $-$ (sum of away team player ratings for players on court)/48. The 3.2/48 represents the average number of points per minute by which the home team defeats the away team. My colleague, Jeff Sagarin of *USA Today*, wrote a program (called WINVAL, [winning value]) to solve for the set of player ratings that best match the scores of the game. Essentially we use a trial set of ratings to predict the score margin for each time segment and adjust the ratings until the set of ratings makes the most accurate predictions possible (aggregated over 38,000 rows of data).

To illustrate how Adjusted $+/-$ differs from Pure $+/-$ let's look at a simplified example. Team 1 has played 20 full-length (48-minute) games against Team 2. Team 1 consists of players 1–9 and Team 2 players 10–18. The results of the games (from the standpoint of team 1) are shown in figure 30.1.

Thus in the first game Players 4, 1, 7, 5, and 2 lost by 13 points to Players 15, 16, 10, 17 and 14. The Pure $+/-$ and Adjusted $+/-$ for these data are shown in figure 30.2. To see that the Adjusted $+/-$ are right for this example note that the predicted score for each game is the

(sum of player ratings for Team 1 players in game)
$-$ (sum of player ratings for Team 2 players in game).

	F	G	H	I	J	K	L	M	N	O	P	Q
3	Game	Result	P 1	P2	P3	P4	P5	P6	P7	P8	P9	P10
4	1	−13	4	1	7	5	2	15	16	10	17	14
5	2	19	1	6	2	5	4	11	17	14	15	18
6	3	−4	1	9	2	8	4	15	14	10	17	13
7	4	29	1	6	5	3	2	16	17	18	14	11
8	5	−3	9	7	1	5	6	17	15	12	18	10
9	6	12	7	2	5	1	4	17	11	15	16	18
10	7	−5	6	5	8	9	1	13	16	12	15	10
11	8	−32	4	2	9	5	3	17	12	10	18	15
12	9	18	8	3	9	1	7	17	16	15	14	11
13	10	17	1	2	9	6	4	13	16	10	11	18
14	11	−11	7	3	2	5	6	14	17	15	12	15
15	12	−14	7	8	4	6	3	18	11	12	17	15
16	13	29	4	5	9	2	6	11	13	14	17	18
17	14	17	1	8	4	2	7	13	12	14	17	18
18	15	0	6	9	8	7	10	15	12	10	17	14
19	16	−7	6	3	2	1	8	17	18	16	14	10
20	17	9	3	2	5	6	7	13	16	14	10	11
21	18	24	1	7	6	7	4	18	13	18	15	11
22	19	18	1	2	5	8	6	14	13	12	15	18
23	20	−24	2	4	3	8	5	11	18	16	17	10

Figure 30.1. Sample data for Adjusted +/−calculations. See file
Newadjustedplusminusex.xls.

	A	B	C
2		Mean: −1.45661E−13	
3			
4	Player	Adjusted +/−	Pure +/−
5	1	12.42497415	8.428571
6	2	2.425071482	4.214286
7	3	−6.575047563	−4
8	4	−10.575047563	2.818182
9	5	−0.575004084	2.333333
10	6	0.424985887	7.769231
11	7	1.425023952	5.363636
12	8	−6.574961831	−0.555556
13	9	5.425007418	2
14	10	16.42495706	6.6
15	11	−13.57506636	−11.9
16	12	0.425010347	4.25
17	13	−9.574964372	−13.125
18	14	−6.574989062	−8.333333
19	15	7.424986222	0.428571
20	16	1.425046331	−4
21	17	−0.574971597	−0.8
22	18	−1.575004236	−7.785714

Figure 30.2. Pure and Adjusted +/−ratings. Note that for each
player, points rating = offense rating − defense rating.

For each game, this equation perfectly predicts the score of the game; thus, these Adjusted +/− ratings must be correct. For example, in Game 1

$$\text{Team 1 total ratings} = -10.58 + 12.42 + 1.43 - .58 + 2.43 = 5.1$$

and

$$\text{Team 2 total ratings} = 7.42 + 1.43 + 16.42 - .57 - 6.57 = 18.1.$$

Thus, we would predict Team 1 to win by $5.1 - 18.1 = -13$ points and this is exactly what happened.

How did we find the Adjusted +/− ? We used the EXCEL SOLVER to choose each player's Adjusted +/− rating to minimize

$$\sum_{i=1}^{i=20} (\text{Sum of Team 1 player ratings for game } i$$
$$- \text{Sum of Team 2 Player ratings for game}$$
$$- \text{Points team 1 wins game } i \text{ by})^2.$$

(See the chapter appendix for an explanation of the Excel Solver.) This is simply the sum of our squared prediction errors over all games. The set of Adjusted +/− in figure 30.2 makes this sum of errors equal to 0. In most cases it is impossible, of course, to get a zero sum of squared errors, but the Solver can easily minimize the sum of squared errors.

Note that Player 4's Adjusted +/− is much worse than his Pure +/− while for Player 10 his Adjusted +/− is much better than his Pure +/− . Let's explain the first anomaly. When Player 4 was on court his team won by an average of 2.82 points. Averaging the ability of his teammates, we find he played with players who on average totaled to 8.34 points better than average. Player 4 played against opponents' lineups that averaged 5.06 points worse than average. Thus, ignoring Player 4, we would have predicted Team 1 to win by $8.34 + 5.06 = 13.4$ points per game. Since Team 1 won by an average of only 2.81 points, we would estimate Player 4's rating as $2.82 - 13.4 = -10.58$.

In our simple example, every game was predicted perfectly by this simple model: Team 1 wins by (sum of Team 1 player ratings for players on court) − (sum of Team 2 player ratings on court). For real data, there is lots of variability and no set of ratings will come close to matching the margins for our 38,000 time segments. Still, there is a set of ratings that best fits the scores, which is what the WINVAL program generates.

2006–2007 Adjusted +/− Ratings

Table 30.2 lists the 25 best-rated players for the 2006–7 season (among players averaging at least twenty minutes per game). Each player's offensive ability and defensive ability have been included in the ratings system. Kevin Garnett's point rating of 19 means that after adjusting for who he played with and against, Garnett would improve the team's performance

TABLE 30.2
WINVAL Top 20 Players' Points Rating, 2006–7

Player	Rank	Points	Offense	Defense
Kevin Garnett	1	19	7	−12
LeBron James	2	15.1	7.6	−7.6
Tim Duncan	3	15.1	5.4	−9.7
Gilbert Arenas	4	13.3	15.8	2.5
Kobe Bryant	5	10.8	7	−3.8
Baron Davis	6	10.7	6.9	−3.8
Paul Pierce	7	10	11.6	1.6
Luol Deng	8	9.6	5.2	−4.4
Carlos Boozer	9	8.9	8.5	−.4
Anthony Parker	10	8.9	6.4	−2.5
Manu Ginobili	11	8.7	10.4	1.7
Elton Brand	12	8.2	3.8	−4.4
Chauncey Billups	13	7.8	11.2	3.4
Dirk Nowitzki	14	7.7	8.1	.4
Steve Nash	15	7.7	8.8	1.1
Marcus Camby	16	7.7	3.7	−4
Jason Kidd	17	7.7	8.2	.5
Jermaine O'Neal	18	7.6	−3.3	−11
Rashard Lewis	19	7.5	4.1	−3.4
Antonio McDyess	20	7.1	−3.1	−10.1

by 19 points per 48 minutes if he played an entire game instead of an average NBA player. Note that only seven players are at least 10 points better than an average NBA player. Garnett has an offense rating of $+7$. This means that after adjusting for who he played with and against, replacing an average NBA player with Garnett would result in the team's scoring 7 more points per game. Also note that (unlike what we found using Pure $+/-$) six of the top twenty players played for teams with a record of .500 or worse (Paul Pierce, Gilbert Arenas, Elton Brand, Jermaine O'Neal, Jason Kidd, and Rashard Lewis). A good offense rating can be created by doing many things: scoring points, getting rebounds, throwing good passes, setting screens, reducing turnovers, and so forth. We do not know how Garnett creates the points (we leave this to the coaches), but we can show how much he helps the offense. Garnett has a defense rating of -12 points. While a positive offense rating is good, a negative defense rating is good. Thus when Garnett replaces an average NBA defensive player, he causes the team to give up 12 fewer points. A good defensive rating can be created by doing many things: blocking shots, stopping the pick and roll, reducing turnovers, causing turnovers, rebounding, and so forth. The beautiful thing about a WINVAL player points rating is that offensive ability and defensive ability are weighted equally. Clearly offense and defense play an equal role in basketball, so this is reasonable. Since NBA box scores calculate many more offensive statistics than defensive statistics, NBA Efficiency and PER and Win Score ratings cannot help but be biased toward great offensive players. In an effort to include a larger defensive component, Win Score is adjusted based on team defense statistics. Unfortunately, such an adjustment fails to account for the fact that Bruce Bowen and Tim Duncan (not Tony Parker and Manu Ginobili) are the keys to the Spurs' defense.

As a byproduct of player ratings WINVAL rates teams. The beauty of the WINVAL team rating system is that since we know who is on the court we can incorporate knowledge of the strengths of opponents faced (rather than just team strength) to rate teams. WINVAL player ratings weighted by the number of minutes played always average out to the WINVAL team rating. Any valid rating system must have this property. To illustrate this, table 30.3 indicates that the average Spurs player had a rating of 1.62. We had the Spurs' team rating as 8.11, which is almost exactly $5 \times (1.62) = 8.1$.

There are several paths to basketball excellence:

- You can be great on both ends of the court (e.g., Kevin Garnett, LeBron James, Tim Duncan, Elton Brand, Luol Deng).
- You can be an offensive star but nothing special on defense (e.g., Gilbert Arenas, Chauncey Billups, Steve Nash, Dirk Nowitzki).
- You can be a defensive star but nothing special on offense (e.g., Jermaine O'Neal or Antonio McDyess).

TABLE 30.3
Minutes Played and Adjusted +/−Player Ratings, Spurs (2006–7)

	D	E	F
8		Minutes	Points
9	Duncan	3462	15.1
10	Manu	2663	8.7
11	Bowen	3153	1.9
12	Barry	1856	−0.7
13	Parker	3249	−3.7
14	Oberto	1780	−0.8
15	Finley	2361	−3.7
16	Vaughn	968	1.3
17	Horry	1484	0
18	Elson	1563	−3.7
19	Bonner	678	−5.6
20	Udrih	968	−5.8
21	Ely	65	−17.8
22	Butler	103	−24.3
23	White	136	−2.8
24	Williams	88	−0.2
25			
26	Player Mean	1.625935	

Most people consider Jermaine O'Neal an excellent offensive player; he averaged over 19 points a game and had a PER rating of 18.8. WINVAL tells us, however, that the Indiana Pacers scored 3.3 points less per game when O'Neal was in the game than they would have scored if an average NBA player replaced him. (I believe this is because the Pacers directed their offense too much toward O'Neal, so that other players failed to get into the flow of the game.)

TABLE 30.4
WINVAL Top 20 Players, Offense Rating,
2006–7

Player	Rank	Points
Gilbert Arenas	1	15.8
Paul Pierce	2	11.6
Chauncey Billups	3	11.2
Andre Miller	4	10.6
Mike Bibby	5	10.6
Manu Ginobili	6	8.7
Dwyane Wade	7	10
Chucky Atkins	8	9.1
Michael Redd	9	8.8
Steve Nash	10	7.7
Ben Gordon	11	8.8
Carlos Boozer	12	8.5
Jason Kidd	13	8.2
Dirk Nowitzki	14	8.1
LeBron James	15	15.1
Chris Paul	16	7.5
Kobe Bryant	17	7
Kevin Garnett	18	7
Baron Davis	19	6.9
Ray Allen	20	6.6

TABLE 30.5
WINVAL Top 20 Players, Defense Rating,
2006–7

Player	Rank	Points
Kevin Garnett	1	−12
Jermaine O'Neal	2	−10.9
Antonio McDyess	3	−10.1
Zydrunas Ilgauskas	4	−10
Tim Duncan	5	−9.7
Jason Collins	6	−8.9
Samuel Dalembert	7	−8.6
Al Jefferson	8	−8.5
Rasheed Wallace	9	−7.9
Nene	10	−7.8
Bruce Bowen	11	−7.7
LeBron James	12	−7.5
Brendan Haywood	13	−7
P. J. Brown	14	−6.9
Nick Collison	15	−6.4
Kendrick Perkins	16	−6.1
Tim Thomas	17	−6.1
Brad Miller	18	−6
Andris Biedrins	19	−5.9
Emeka Okafor	20	−5.4

Table 30.4 shows the top twenty players for offense rating; table 30.5 shows the top twenty players for defense rating. Note that seventeen of the top twenty offense ratings belong to guards, with Nowitzki, Boozer, and Garnett as exceptions. Also note that the only guard in the top twenty defenders is LeBron James. (This is reasonable because "inside" players often help out on defense and alter or block shots.) Devin Harris is the highest-rated point guard on defense (#25 with −4.8 rating).

Let's examine further several NBA players to better explain their ratings.

Why Is Kevin Garnett the Best?

Win Score indicates that Kevin Garnett is the best NBA player of the last ten years (I agree). To make the case for Garnett's greatness during the 2006–7 season, let's look at the following data, which indicate how the 2006–7 Minnesota Timberwolves played when Garnett was in with other players, and when he was in and other players were out.

For example, with Garnett in and Mike James in, the Timberwolves were 1.1 points better per 48 minutes than an average team, but with Garnett out and James in, the team was 25.2 points worse per 48 minutes than an average NBA team. Figure 30.3 shows that for every player (except Justin Reed, who played little), taking Garnett out leads to an unmitigated disaster.

	D	E	F
8		**Garnett in**	**Garnett out**
9	Garnett	2.068623241	dnp
10	Smith	6.625117829	−19.1260921
11	Foye	2.687658381	−18.8363539
12	Jaric	1.528346916	−17.4928541
13	Davis	2.062763933	−19.804282
14	James	1.149827728	−25.2165387
15	Hassell	1.741019578	−25.8873062
16	Blount	1.1715841	−21.7697961
17	Madsen	6.752480906	−14.093463
18	Hudson	−1.458252856	−15.0936761
19	McCants	12.04746088	−27.8568458
20	Reed	−20.23843887	−11.1550147

Figure 30.3. The Greatness That Is Kevin Garnett. *Note:* dnp = did not play.

Proof That Anthony Parker Played Well in 2006–7

Anthony Parker played in Europe for several years and joined the Toronto Raptors as their starting point guard for the 2006–7 season. The Raptors had a great year (47–35 record) and won their division. Chris Bosh and Coach Sam Mitchell received most of the credit for the Raptors' improvement. However, Anthony Parker should have received some credit for the Raptors' surprising success. Parker's WINVAL points rating for 2006–7 of 8.9 was tenth best in the league. Chris Bosh had a +6 points rating (#33 in the league). According to WINVAL, Parker contributed around 9 points per 48 minutes more to the Raptors' success

	BB	BC	BD	BE	BF
13		**Bosh in**	**Parker in**	**Bosh out**	**Parker out**
14	Bargnani	2.9675406	7.0531174	−5.080627	−9.551004043
15	Bosh	3.4984239	6.7634537	dnp	−4.352483864
16	Calderon	3.259113	11.07591	−7.232144	−13.40031441
17	Parker	6.7634537	5.5126843	1.5550114	dnp
18	Peterson	−0.6126	3.9678528	−9.446316	−10.4388624
19	Graham	6.6900567	9.9500844	−8.07867	−10.66379074
20	Ford	3.0776533	2.6288383	−5.047972	−2.634105994
21	Dixon	1.2879943	11.21127	−14.18934	−10.42451286
22	Nesterovic	3.5334955	4.7620006	1.988912	−2.173419271
23	Garbajosa	6.9168003	4.1826265	−6.915919	−2.076643361
24	Jones	−4.702635	−3.06935	−14.38016	−14.50744523
25	Jackson	10.044726	25.429229	−8.665299	−14.25143233

Figure 30.4. Anthony Parker Played Well during the 2006−7 Season. *Note:* dnp = did not play.

than an average NBA player would have. Parker's PER rating was 14.4 (near average) and Bosh's PER rating was 22.6. Parker's Win Score per minute was 0.18 and Bosh's Win Score per minute was 0.28. An average NBA player has a Win Score around 0.17. Thus, according to both PER and Win Score, Bosh is fantastic and Parker is below average (according to PER) and around average (according to Win Score). Let's look at how the Raptors played in 2006–7 with Parker on and off the floor.

Figure 30.4 shows that except when Fred Jones was in, the Raptors played well when Parker was in. When Parker was out, nothing worked. Even with Chris Bosh in the Raptors played at a −4.35 level. With Bosh out and Parker in, the Raptors still played at an above-average level (+1.55).

Is Kevin Martin a Great Player?

Kevin Martin of the Sacramento Kings was the runner-up to Monta Ellis of the Golden State Warriors in the voting for the 2006–7 Most Improved Player award. Martin's 2006–7 PER rating was 20.1 (this is well above average; fewer than thirty players had a PER rating above 20) and a Win Score of 0.21 (again, well above average). WINVAL gives Martin a points rating of −3.4 (#155 out of 200 regulars), an offense rating of 5.5 (#26 out of 200 regulars) and a defense rating of 8.9 (#198 out of 200 regulars). Thus, Martin's WINVAL score indicates he hurts the team while PER and Win Score indicate that Martin is a well above average NBA player. Since PER and Win Score overemphasize a player's offensive contribution and Martin is a great offensive player, it is clear why PER and Win Score indicate Martin is a great

	V	W	X	Y
7		**Martin in**	**Martin out**	**Out / In**
8	Artest	1.520169	3.73181232	2.21164324
9	Douby	−13.3726	−9.9259229	3.44670514
10	Garcia	−0.55863	−3.94411	−3.385484
11	Salmons	−5.52131	−2.6110906	2.91021975
12	Williams	−22.1789	−0.9631193	21.2158066
13	Abdur-Rahim	−3.33606	−3.1297542	0.20630842
14	Bibby	0.44753	4.52107449	4.07354471
15	Martin	−0.86647	dnp	not relevant
16	Miller	6.129669	−7.32	−13.45
17	Price	−9.03113	−5.3525623	3.67856876
18	Williamson	−1.01812	4.88507185	5.90318911
19	Thomas	−2.93184	−12.229373	−9.2975287

Figure 30.5. Kevin Martin is overrated.

player. Figure 30.5 shows how the 2006–7 Kings performed with Martin in and out of the game. The Kings were better when Martin was in with Miller than with Miller in and Martin out. This may indicate that Miller's good defense erased some of Martin's defensive shortcomings. But for key Kings players like Ron Artest, Mike Bibby, Corliss Williamson, and John Salmons, the Kings were worse with Martin in than with Martin out. These data provide little evidence that Martin is one of the league's best players.

Why Does Jason Collins Have a Horrible PER Rating?

In 2003–4 Jason Collins of the New Jersey Nets had a PER rating of 10, and in 2004–5 it fell to 8.6. Both ratings are well below the league average of 15. In 2006–7, Collins had a horrible Win Score of 0.07. Despite these poor metrics, many coaches think Collins is an excellent defender. Table 30.6

TABLE 30.6
Jason Collins Is a Great Defender

Year	Points Rating	Offense Rating	Defense Rating
2003–4	3.4	−5.3	−8.7
2004–5	4.1	−8.3	−12.4
2005–6	−3.3	−9	−5.7
2006–7	−0.4	−9.6	−9.2

shows Collins's great defensive contributions. WINVAL paints a consistent picture of Jason Collins's abilities. He really hurts the Nets' offense, but he is a fantastic defender, always ranking in the league's top twenty-five and usually in the top ten.

Impact Ratings

Throughout this chapter I have sung the praises of WINVAL. So what are its shortcomings? First, there is a lot of "noise in the system." It takes many minutes to get an accurate player rating. I do not have a great deal of confidence in WINVAL ratings for a player who plays less than 500 minutes in a season. Second, you can accumulate your "good or bad" statistics at times when the game is decided. For example, suppose I enter the game with three minutes to go and my team is down by 20 points. The game outcome is certainly decided at this point. Yet if my team cuts the opponent's lead to 3 points by the end of the game I will pick up a lot of credit toward my WINVAL point rating. To remedy this problem, WINVAL has created the WINVAL Impact rating. The WINVAL Impact rating is similar to the SAGWIN points for baseball described in chapter 8. I examined thousands of NBA game scores as the game progresses and determined the probability (if two teams of equal ability play) of a team winning given the time left in the game and the score of the game. To create the WINVAL Impact rating we create an alternate scoreboard where the score at any point in the game is the team's chance of winning the game. Then a player gains credit for the change in the team's chance of winning, instead of the change in the score of the game. For example, at the start of the game our alternate scoreboard has a score of 50–50. If I play for five minutes and the score after five minutes is 14–5 in favor of my team, the regular scoreboard says we have picked up 9 points (which feeds into my point rating), while the Impact scoreboard says that my chance of winning is now 72–28, so my team has picked up 22 impact points. Suppose we are down by 5 with two minutes to go. The alternate scoreboard says we are down 11–89. If we win the game my team has picked up 89 Impact points. We then use the change in the alternate scoreboard as the input to our Impact rating analysis. Table 30.7 lists the top ten Impact ratings among regulars for the 2006–7 season.

We interpret a player's Impact rating as follows. A player with an Impact rating (expressed as a decimal) of x would win a fraction of $0.5 + x$ of his games if he played with four average NBA players against a team of five

TABLE 30.7
Top 10 Impact Ratings, 2006–7 Season

Player	Rank	Impact (%)
Kevin Garnett	1	42
LeBron James	2	34
Anthony Parker	3	31
Chauncey Billups	4	31
Paul Pierce	5	31
Gilbert Arenas	6	30
Tim Duncan	7	29
Chris Paul	8	26
Baron Davis	9	26
Elton Brand	10	25

average NBA players. For example, if Kevin Garnett playing with average guys would beat five average guys 92% of the time, while LeBron James playing with four average guys would beat a team of five average guys 84% of the time.

Breaking down each player's rating by quarter allows us to measure a player's clutch ability using his fourth-quarter rating. Table 30.8 gives the top ten fourth-quarter performers for the 2006–7 season. LeBron James's fourth-quarter rating of $+20$, for example, means the Cleveland Cavaliers play 20 points (per 48 minutes) better during the fourth quarter than they would if an average NBA player replaced James.

The Roland Ratings from 82games.com

In recent years the Web site www.82games.com has included player ratings. This Web site gives many great statistics for players and teams. For example, for each player the site provides EFG for jump shots and close-in shots. Figure 30.6 lists the 82games.com ratings for the 2006–7 champion San Antonio Spurs.

The Roland ratings seem to be an admirable effort to combine the Linear Weights and Adjusted $+/-$ models. If the Roland player ratings are accurate, then the average rating of the players on a team weighted by the minutes

TABLE 30.8
Top 10 Fourth-Quarter Adjusted + / − for 2006–7 Season

Player	Rank	Fourth-Quarter Points Rating
LeBron James	1	20
Kevin Garnett	2	18.8
Carlos Boozer	3	15.6
Paul Pierce	4	14.1
Anthony Parker	5	13.4
Mike Miller	6	12.8
Chauncey Billups	7	11.9
Michael Redd	8	11.2
Chucky Atkins	9	11
Kobe Bryant	10	10.6

should approximate a measure of the team's actual ability level. Most sports analysts believe the Spurs were 7–8 points better than average during the 2006–7 season.[3] The weighted average by minutes of the Roland ratings for Spurs' players estimates that the Spurs are 13.8 points better than average. Even the weighted average of the net for on court versus off court indicates the Spurs are 10.85 points better than average. Some adjustment needs to be made to the Roland ratings to ensure that the player ratings are consistent with the team's overall ability.

How Well Did Tony Parker Play during 2006–7?

How can we justify Tony Parker's −3.8 points rating? He was MVP of the NBA Finals and Eva Longoria loves him! Table 30.9 demonstrates why Tony Parker did not play as well as people think he did during 2006–7. In many situations, the Spurs played better with Parker out than with Parker in. For example, when Duncan, Ginobili, and Bowen were in and Parker was in the Spurs played at a +16 level, while when Duncan, Ginobili, and Bowen were in and Parker was out the Spurs played at a +28 level.

[3] See, for example, the Sagarin ratings on http://www.usatoday.com/sports/sagarin.htm.

	H	I	J	K	L	M	N	O	P
5		**Production**				**On Court/Off Court**			**Roland**
6	**Player**	**Min.**	**Own**	**Opp.**	**Net**	**On**	**Off**	**Net**	**Rating**
7	Duncan	0.69	28.4	13.6	14.8	13	−2	15	14.9
8	Ginobili	0.52	26.3	11.7	14.5	12.6	3.7	8.9	12.9
9	Parker	0.63	23.4	13.7	9.7	9.8	5.8	4	8.1
10	Barry	0.41	18.1	12.9	5.2	6.4	9.7	−3.3	2.8
11	Bonner	0.16	18	13.1	4.9	2.1	9.6	−7.5	1.4
12	Bowen	0.62	7.9	13.7	−5.8	11.5	3.2	8.3	−1.8
13	Finley	0.46	15.1	15.2	−0.1	4.7	11.5	−6.8	−2.1
14	Horry	0.28	13.1	16.7	−3.6	8.3	8.4	−0.1	−2.6
15	White	0.03	13.4	11.3	2.1	−7.7	8.9	−16.7	−3.2
16	Oberto	0.34	13.2	17.6	−4.5	7.2	9	−1.8	−3.7
17	Vaughn	0.19	12.1	18.3	−6.2	10	8	2	−3.9
18	Udrih	0.24	11.4	14.6	−3.2	1	10.7	−9.7	−5
19	Williams	0.02	16.3	21.7	−5.3	3.3	8.5	−5.2	−5.3
20	Elson	0.34	12.3	20.6	−8.2	5.2	10	−4.8	−7.2
21	Ely	0.02	7.6	20.5	−12.9	−19.9	8.8	−28.8	−17.4
22	Butler	0.03	10.7	30.4	−19.7	−17.7	9.1	−26.8	−21.8

Figure 30.6. San Antonio Spurs' ratings for the 2006–season. Min = fraction of team minutes that a player was on court. Thus Tim Duncan was on court 69% of the time. Production Own: This is the player's PER rating when on court. Thus Duncan had a 28.4 PER rating. Production Opp: This is the PER rating for the opposing team's player who is playing the same position as the player. For example, Duncan played some time at power forward and at center. The PER for his counterpart on the opposing team was 13.6. Production Net: This is simply Production Own−Production Opp. For Duncan this is 28.4 − 13.6 = 14.8 On: This is the player's Pure +/− per 48 minutes. Thus when Duncan was on the court the Spurs won by 13 points per 48 minutes. Off: This is the number of points by which the Spurs outscored opponents with the player off the court (per 48 minutes). With Duncan off the court the Spurs lost by an average of 2 points per game. Net On/Off: This is simply On–Off and is a crude estimate of the player's Adjusted +/− rating. Thus net for Tim Duncan = 13−(−2) = 15, which is close to our Adjusted +/−. For Tony Parker our Adjusted +/−is −3.7, while the net =4. The difference here is because Parker usually played with outstanding players like Duncan and Ginobili. This will inflate his Pure +/− number and give us an inflated view of his ability. Roland rating: This seems to be .72 × (net for PER) + .28 × (net for On/Off). Thus Tony Parker's Roland rating is .72(9.7) + .28(4) = 8.1. Source: http://www.82games.com.

Measuring Player Improvement

The WINVAL ratings extend back through the 1998–99 season, which allows us to study a player's development over time. For example, table 30.10 shows LeBron James's ratings by season. James was nothing special as a rookie, but his continuous improvement bodes well for the Cavs and should scare the rest of the NBA.

TABLE 30.9
Spurs' Rating, 2006–7

Other Players	Parker in	Parker out
Duncan, Ginobili, Bowen in	+16	+28
Duncan, Ginobili in	+16	+11
Ginobili, Bowen in	+15	+16
Duncan, Bowen in	+12	+21
Bowen, Duncan in; Ginobili out	+9	+14

TABLE 30.10
Points Rating for LeBron James

Year	Rating
2003–4	.9
2004–5	7.9
2005–6	9.5
2006–7	15.1

We can also use WINVAL to summarize a player's ratings by month during the regular season and playoffs. Figure 30.7 shows Dirk Nowitzki's monthly point ratings during his MVP–winning 2006–7 season. As the regular season drew to a close Nowitzki's performance declined. This turned out to be a harbinger of his (and the Mavericks') poor playoff performance against the Golden State Warriors.

As another example of the insights that can be gleaned from the monthly ratings, consider DeShawn Stevenson of the Utah Jazz during the 2002–3 season. His points rating during the 2002–3 season was −7, but his points rating dropped to −22 during February through March. In April 2003 he was suspended (for unknown reasons) by Jazz coach Jerry Sloan. His performance decline indicated that he was having trouble playing at an acceptable level within Sloan's system, and during the next season he was traded to the Orlando Magic.

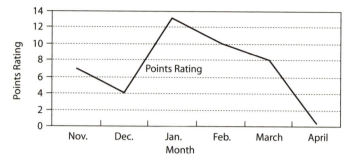

Figure 30.7. Dirk Nowitzki's monthly point ratings for the 2006−7 season.

During the season Jeff Sagarin and I tabulate a listing of the "hot" players based on each team's last five games. When the Mavericks play a team this aids them in setting up a game plan: concentrate on the hot players and "go at" the weak players (particularly on defense).[4]

APPENDIX

Using the Excel Solver to Find Adjusted +/−Ratings

To install the Excel Solver in Excel 2003 or an earlier version of Excel simply select Tools followed by Add-ins and then check the Solver box and hit OK. In Excel 2007 you can install the Solver by selecting the Office Button followed by Add-ins. Then choose Go and check the Solver Add-in box and select OK.

You may now activate the Excel Solver in Excel 2003 or earlier by selecting the Solver from the Tools menu. In Excel 2007 you may now activate the Solver by selecting the Solver from the right-hand portion of the data tab. When you activate the Solver you will see a dialog box (figure 30.8).

There are three parts to a Solver model:

- Target Cell: This is an objective you can maximize or minimize. In our example our objective or Target Cell is to minimize the sum of the squared prediction errors for each game.
- Changing Cells: These are the cells in the spreadsheet that Solver is allowed to change to optimize the Target Cell. In our example the Changing Cells are the Adjusted +/− ratings for each player.

---

[4] Jeff Sagarin and I are consultants to the Dallas Mavericks.

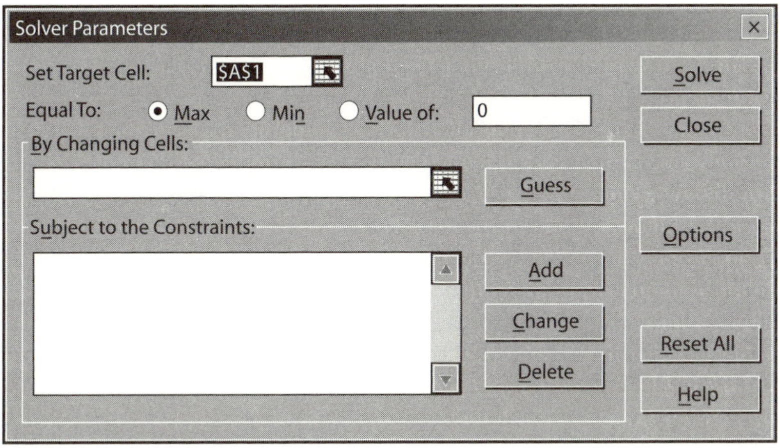

Figure 30.8. Dialog box.

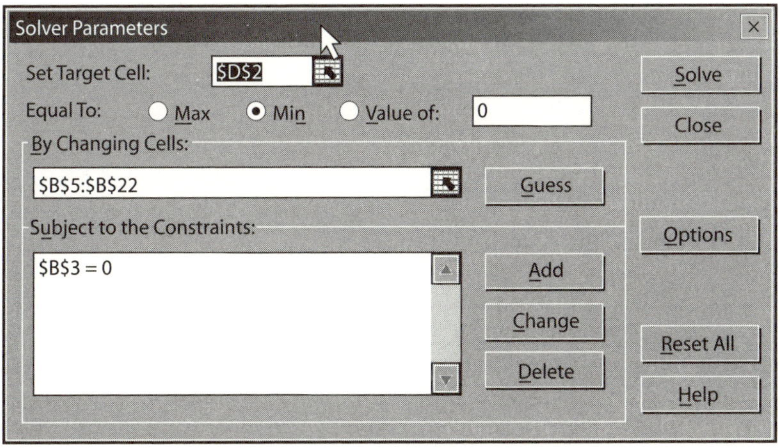

Figure 30.9. Solver window for computing Adjusted +/− ratings. See file Newadjustedplusminusex.xls.

- Constraints: Constraints are restrictions on the Changing Cells. In our example the only constraint we included was to make the average player rating (weighted by games played) equal to 0. This ensures that an above-average player will have a positive rating and a below-average player will have a negative rating.

Figure 30.10 shows the setup of the spreadsheet for determining Adjusted +/− ratings. In column B we begin by entering any trial set of

	A	B	C	D	E	F	G
1				**SSE**	**Player**		
2		**Mean**		1.03053E−08	4		
3		−6.29126E−15		**Squared Error**	**Prediction**	**Game**	**Result**
4	**Player**	**Adjusted +/−**	**Pure +/−**	6.81E−10	−13.00001946	1	−13
5	1	12.42497415	8.428571	9.64E−11	19.00001757	2	19
6	2	2.425071482	4.214286	1.06E−10	−3.99997432	3	−4
7	3	−6.575047563	−4	1.24E−09	28.99996428	4	29
8	4	−10.57506265	2.818182	9.07E−11	−2.99997899	5	−3
9	5	−0.575004084	2.333333	1.56E−10	11.99999458	6	12
10	6	0.424985887	7.769231	1.16E−09	−5.00003072	7	−5
11	7	1.425023952	5.363636	1.74E−10	−32.0000105	8	−32
12	8	−6.574961831	−0.555556	8.86E−11	18.00000071	9	18
13	9	5.425007418	2	6.18E−11	16.99999218	10	17
14	10	16.42495706	6.6	5.69E−11	−10.9999974	11	−11
15	11	−13.57506636	−11.9	2.75E−10	−14.0000129	12	−14
16	12	0.425010347	4.25	3.99E−11	28.99999081	13	29
17	13	−9.574964372	−13.125	1.29E−09	16.99995494	14	17
18	14	−6.574989062	−8.333333	1.34E−09	1.48057E−05	15	0
19	15	7.424986222	0.428571	2.68E−10	−7.00000523	16	−7
20	16	1.425046331	−4	2.12E−09	9.00004219	17	9
21	17	−0.574971597	−0.8	2.98E−12	24.00000893	18	24
22	18	−1.575004236	−7.785714	7.13E−10	18.0000229	19	18
23				1.17E−09	−23.9999657	20	−24

Figure 30.10. Solver setup for computing Adjusted +/− ratings.

Adjusted +/− ratings. These ratings are averaged in cell B3. In column S we use lookup functions to determine the sum of the Adjusted +/− for our team during each game. In column T we use lookup functions to determine the sum of the Adjusted +/− for our opponents during each game. Next in column E we created our "prediction" for each game by taking the (sum of our player rating) [from column S] − (sum of opponent player ratings) [from column T]. Then in column D we computed the squared error for each of our game predictions by computing (column G − column E)^2. In cell D2 we computed the sum of squared prediction errors for each game.

The Solver Window needed to compute our Adjusted +/− ratings is shown in figure 30.9. Simply minimize the sum of squared errors (cell D2) by changing the Adjusted +/− ratings (cells B5:B22). Then constrain the average rating (computed in cell B3) to equal 0. Just click Solve and you have found the Adjusted +/− ratings that most accurately predict the game scores.

	H	I	J	K	L	M	N	O	P	Q	R	S	T
3	P1	P2	P3	P4	P5	P6	P7	P8	P9	P10	In?	Us	Them
4	4	1	7	5	2	15	16	10	17	14	1	5.125	18.1
5	1	6	2	5	4	11	17	14	15	18	1	4.125	−14.9
6	1	9	2	8	4	15	14	10	17	13	1	3.125	7.13
7	1	6	5	3	2	16	17	18	14	11	0	8.125	−20.9
8	9	7	1	5	6	17	15	12	18	10	0	19.12	22.1
9	7	2	5	1	4	17	11	15	16	18	1	5.125	−6.88
10	6	5	8	9	1	13	16	12	15	10	0	11.13	16.1
11	4	2	9	5	3	17	12	10	18	15	1	−9.875	22.1
12	8	3	9	1	7	17	16	15	14	11	0	6.125	−11.9
13	1	2	9	6	4	13	16	10	11	18	1	10.12	−6.88
14	7	3	2	5	6	14	17	15	12	15	0	−2.875	8.13
15	7	8	4	6	3	18	11	12	17	15	1	−21.88	−7.88
16	4	5	9	2	6	11	13	14	17	18	1	−2.875	−31.9
17	1	8	4	2	7	13	12	14	17	18	1	−0.875	−17.9
18	6	9	8	7	10	15	12	10	17	14	0	13.13	17.1
19	6	3	2	1	8	17	18	16	14	10	0	2.125	9.13
20	3	2	5	6	7	13	16	14	10	11	0	−2.875	−11.9
21	1	7	6	7	4	18	13	18	15	11	1	5.125	−18.9
22	1	2	5	8	6	14	13	12	15	18	0	8.125	−9.87
23	2	4	3	8	5	11	18	16	17	10	1	−21.88	2.12

Figure 30.10. (*cont.*)

If the games are not of equal length, then for each "game" you should predict the point differential per minute and then compute the squared prediction error for each game on a per-minute basis. To compute the Target Cell you weight each game's per-minute squared error by the number of minutes in the game.

NBA LINEUP ANALYSIS

In chapter 30 we described the methodology for creating Adjusted +/− ratings. These are helpful to teams in making decisions involving players such as trades and salaries. During the season, however, few players are traded and a team's major concern is how to win more games with their current roster. The most important decisions coaches make during the season are which lineups to play when. For example, should a team try to go small against the 2006–7 Suns' "small ball," or should they go big and push the ball inside?

On average a team plays 300–600 different lineups during the course of a season. Is there any rhyme or reason to coaches' lineup choices? Lineup ratings can help a team win more games: play better lineups more and worse lineups less.

Once we have player ratings it is easy to develop lineup ratings. Suppose we want to rate the Indiana Pacers' lineup of Jermaine O'Neal, Danny Granger, Jamaal Tinsley, Troy Murphy, and Mike Dunleavy. Let's call this lineup Pacers 1A. During the 2006–7 season this lineup played 326 minutes (more than any other Pacer lineup). They outscored their opponents by 11 points. This means that Pacers 1A played $326/48 = 6.79$ games and has a Pure +/− of $11/6.79 = 1.61$ points per 48 minutes. Then we look at each minute Pacer 1A was on the court and average the total abilities of the opponents (adjusting, of course, for the league home edge of 3.2 points). We find this lineup played against opposition lineups averaging in ability 1.89 points better than league average. This means that Pacers' 1A should have an Adjusted +/− rating of $1.61 + 1.89 = 3.5$ points. In short, this means that Pacers 1A would beat an average NBA lineup (a lineup where the sum of the five player ratings is 0) by 3.5 points per 48 minutes. The amazing thing is that many teams play inferior lineups many more minutes than they do their better lineups. For example, in 2006–7 the Charlotte Hornets played the lineup of Raymond Felton, Emeka Okafor, Adam Morrison, Brevin Knight, and Gerald Wallace 167 minutes at a −5 level and lost by 44 points to their opponents. This is not a good performance, yet this was the lineup used the most by Charlotte during the season. Substituting Matt Carroll for Brevin Knight in this lineup yields a lineup that played 144 minutes at a +20 level and beat its opponents by 58 points. Why not play the second lineup more often?

The Cleveland Cavaliers played Drew Gooden, Larry Hughes, LeBron James, Eric Snow, and Anderson Varejo 156 minutes during the 2006–7 season. This lineup played at a −14 level and was beaten soundly by 56 points. The lineup resulting from simply substituting Sasha Pavlovic for Eric Snow played 124 minutes and played at a +25 level and outscored the Cavs' opponents by 58 points. Why not play the better lineup more?

For seven years the Dallas Mavericks have factored these lineup ratings into their coaching decisions. Over the course of these seven years the Mavs have won more games than any team except for the Spurs. We cannot prove the link between our lineup ratings and the Mavs' performance, but playing good lineups more and worse lineups less makes perfect sense.

Critics might say that lineups do not play many minutes, so there must be a lot of variability in the lineup ratings. What is the probability that a given lineup is really better than another lineup? The spreadsheet Lineupssuperiority.xls answers this question. Enter the minutes played and lineup rating for the better lineup in row 4 and the same data for the inferior lineup in row 5. Cell E8 indicates the chance that the lineup with the higher rating is truly better. There is over a 99% chance that the Cavs' lineup with Pavlovic is actually superior to the lineup with Snow. This should convince the Cavs that the inferior lineup should play little, if at all.

How did we determine the true probability of lineup 1 being better than lineup 2? Data indicate that the actual performance of a lineup over 48 minutes is normally distributed with a mean equal to the lineup rating and a standard deviation of 12 points. The standard deviation of a lineup's rating is

$$\frac{12}{\sqrt{\text{Games played}}}.$$

For example, for a lineup that plays 192 minutes, our rating has a standard deviation of $12/2 = 6$ points. A basic theorem in statistics tells us the variance of the difference of independent random variables is the sum of the

	B	C	D	E	F
1	**Hughes/Varejao/James/Gooden/5th Player**				
2					
3	**5th Player**	**Lineup #**	**Rating**	**Minutes**	**Games**
4	Pavlovic	1	25	124	2.583333333
5	Snow	2	−14	156	3.25
6					
7	**Difference**	**Variance difference**	**Sigma difference**	**Probability Lineup 1 is better**	
8	39	100.0496278	10.00248108	0.999980064	

Figure 31.1. Lineup superiority calculator.

random variable variances. Since the standard deviation of a random variable is simply the square root of the random variable's variance, the standard deviation of the difference in the rating of two lineups is

$$\sigma = \sqrt{\frac{144}{\text{gameslineup 1 played}} + \frac{144}{\text{gameslineup 2 played}}}.$$

Here games played = minutes played/48.

Then we find the probability that lineup 1 is actually better equals the probability that a random variable with mean lineup 1 rating − lineup 2 rating, and standard deviation σ is greater than 0. This probability may be found with the EXCEL formula

1 − NORMDIST(0,lineup 1 rating − lineup 2 rating, σ, True).

Figure 31.1 shows that there is over a 99% chance that the Pavlovic lineup is better than the Snow lineup.

Lineup Chemistry 101

Coaches often say that a given lineup has great "on-court chemistry" when it really clicks. When a lineup plays poorly coaches say these guys play "like they have never seen each other before." We can easily identify lineups with good or bad chemistry. We define the lineup's chemistry value as (lineup rating) − (sum of individual player ratings). A positive chemistry rating indicates that a team played better than expected and exhibits positive chemistry, while a negative chemistry rating indicates that a team played worse than expected and exhibits negative synergy.

Some examples of team chemistry are shown in figure 31.2. For example, Chauncey Billups, Rick Hamilton, Jason Maxiell, Tayshaun Prince, and Chris

	A	B	C	D	E	F	G
60	**Team**						**Chemistry**
61	**Detroit**	Billups	Hamilton	Maxiell	Prince	Webber	25.6
62	**Golden State**	Barnes	Biedrins	Davis	Ellis	Pietrus	−10.1
63	**Memphis**	Gasol	Jones	Miller	Roberts	Stoudamire	−20.1
64	**Atlanta**	Childress	Johnson	Pachulia	J. Smith	M. Williams	19.6
65	**Boston**	Jefferson	Perkins	Pierce	Scalabrine	West	−16.2
66	**Charlotte**	Carroll	Felton	Morrison	Okafor	G. Wallace	20.2
67	**Chicago**	Deng	Duhon	Hinrich	Thomas	B. Wallace	−21.7
68	**Cleveland**	Hughes	Ilgauskas	James	Pavlovic	Varejao	−23.6
69	**Dallas**	Diop	Harris	Howard	Nowitzki	Stackhouse	27.4
70	**Denver**	Hilario	Kleiza	Najera	J. R. Smith	Iverson	−14.5
71	**Phoenix**	Barbosa	Bell	Diaw	Marion	Stoudemire	18.5
72	**San Antonio**	Bowen	Duncan	Ginobili	Horry	Parker	−9.3

Figure 31.2. Examples of good and bad lineup chemistry.

Webber played 25.6 points better per 48 minutes than expected, while Pau Gasol, Eddie Jones, Mike Miller, Lawrence Roberts, and Damon Stoudamire played 20.1 points worse than expected. It would be helpful to discover what factors lead to positive and negative chemistry. For example, if a lineup has too many scorers and not enough effective defensive players, does that tend to lead to negative lineup chemistry?

Coaches often say how well a lineup plays depends on who the opposition puts on the court. We will analyze matchups in the next chapter.

ANALYZING TEAM AND
INDIVIDUAL MATCHUPS

Among other things, a successful coach must be a master psychologist who can motivate players to play for the team rather than themselves (there is "no *I* in team"; "the whole is greater than the sum of its parts"). A great coach must have sound offensive and defensive strategies and get the players to buy into executing his strategic concepts. Great coaches also have excellent insight as to which players to put in a game at a given time to best match up with the opponent's lineup. Recall from chapter 30 that we have Adjusted $+/-$ ratings for each NBA player. We can also break down each player's Adjusted $+/-$ by opponent. This helps coaches determine how to match their players up with those of their opponents.

Spurs–Mavericks 2006 Western Conference Semifinal

The Dallas Mavericks' surprising journey to the NBA finals during the 2006 playoffs provides a great illustration of how useful the team-by-team breakdown of Adjusted $+/-$ can be. Few analysts gave the Mavs any chance against the world champion Spurs. A key coaching move during this series was the insertion of guard Devin Harris into the starting lineup in place of Adrian Griffin. This surprise decision enabled the Mavs to "steal" game 2 at San Antonio, and the Mavs went on to win the series in seven games by winning an overtime thriller game 7 at San Antonio. Team-by-team Adjusted $+/-$ ratings were an important factor in making this coaching decision. Devin Harris's overall 2005–6 rating was -2.1 points and his Impact rating was -15%. Against the Spurs, Harris's point rating was $+9.4$ points and his Impact rating was $+8\%$. Against the Spurs, Griffin had a -5 point rating and -18% Impact rating. More important, Griffin had a $+18$ points offense rating. This indicates that against the Spurs, Griffin devastated the Mavs' offense. Given these data, it seems clear that Harris should have started in lieu of Griffin. So how did Harris do? During the playoffs we do a two-way lineup calculator, which shows how the Mavs do with any combination of Mavs players on or off court against any combination of the opponent's players on or off court. Doing this calculation for 2005–6 regular season data indicated that Harris could outplay Tony Parker. So what happened during the first six games of the Spurs–Mavs series? With Harris on the court against Parker, the Mavs (excluding game 2, in which the Mavs took Spurs by surprise by starting Harris) beat the Spurs by an average of 102–100 per 48 minutes with Harris in against Parker. With Parker on the floor and Harris out, the Mavs lost by an average of 96–81 per 48 minutes. As the series progressed we learned other interesting things. When Marquis Daniels was on the court against Manu Ginobili, the Mavs lost by an average score of 132–81 per 48 minutes. When Ginobili was on the court and Daniels was out, the Mavs won by an average of 94–91 per 48 minutes. (Daniels did not play in game 7 of the series.)

The Mavs went on to play the Suns in the 2006 Western Conference Finals. Here we knew Devin Harris would not be as effective because he has problems guarding Steve Nash. For this series, when the Harris–Jason Terry backcourt played against Nash, the Mavs were down 113–90 per 48 minutes. The Terry–Jerry Stackhouse backcourt (with Harris out) was up 116–96 per 48 minutes with Nash in. The Mavs used this information to adjust their rotation to play the more effective combination more often.

During a playoff series we also track how well each lineup plays. Table 32.1 shows how the most frequently used lineups performed through five games of the Spurs–Mavs playoff series. For example, lineup 3 for the Mavs (Dampier, Harris, Nowitzki, Stackhouse, and Terry) played 21.34 minutes and was up by 22 points. This lineup played at a +60.55 level. Lineup 7 for the Mavs played very poorly (Dampier, Daniels, Howard, Stackhouse, and Terry). This lineup lost by 14 points in 8.92 minutes and played at a −65.28 level.

The Non-Transitivity of NBA Matchups

In basketball, transitivity would indicate that if Player A is better than Player B and Player B is better than Player C, then Player A is better than Player C. The following discussion shows that transitivity does not hold for basketball player matchups.

We have seen that Devin Harris can outplay Tony Parker and Steve Nash can outplay Devin Harris (thus, transitivity would indicate that Nash can

TABLE 32.1
Spurs–Mavericks 2006 Playoff Lineup Analysis

Dallas Mavericks

						Rating	Minutes	Simple	Raw +/−	Lineup Code#
1	Diop	Harris	Howard	Nowitzki	Terry	16.36	41.04	1.18	1	8816_DAL_2006
2	Dampier	Howard	Nowitzki	Stackhouse	Terry	29.24	31.06	15.45	10	12868_DAL_2006**
3	Dampier	Harris	Nowitzki	Stackhouse	Terry	60.55	21.34	49.47	22	12836_DAL_2006**
4	Dampier	Harris	Howard	Nowitzki	Terry	−1.78	15.89	−18.12	−6	8804_DAL_2006***
5	Dampier	Harris	Howard	Nowitzki	Stackhouse	−20.83	14.19	−33.80	−10	4708_DAL_2006***
6	Diop	Harris	Nowitzki	Stackhouse	Terry	15.54	11.48	4.18	1	12848_DAL_2006
7	Dampier	Daniels	Howard	Stackhouse	Terry	−65.28	8.92	−75.26	−14	12364_DAL_2006***
8	Diop	Howard	Nowitzki	Stackhouse	Terry	−26.13	8.79	−43.62	−8	12880_DAL_2006***
9	Diop	Harris	Howard	Nowitzki	Stackhouse	81.91	8.02	65.92	11	4720_DAL_2006**
10	Daniels	Harris	Howard	Nowitzki	Stackhouse	78.79	7.82	67.55	11	4712_DAL_2006**
11	Diop	Howard	Nowitzki	Terry	Griffin	−9.51	6.92	−34.70	−5	41552_DAL_2006***
12	Harris	Howard	Nowitzki	Stackhouse	Terry	44.11	6.78	35.43	5	12896_DAL_2006**
13	Dampier	Daniels	Nowitzki	Stackhouse	Terry	−42.30	6.37	−52.81	−7	12812_DAL_2006***

San Antonio Spurs

1	Bowen	Duncan	Finley	Ginobili	Parker	21.08	45.83	6.29	6	1054_SAS_2006**
2	Barry	Bowen	Duncan	Finley	Parker	37.90	33.42	20.12	14	1039_SAS_2006**
3	Bowen	Duncan	Ginobili	Horry	Parker	17.94	28.23	5.10	3	1078_SAS_2006
4	Barry	Bowen	Duncan	Horry	Parker	−26.88	10.05	−47.75	−10	1063_SAS_2006***
5	Barry	Duncan	Finley	Ginobili	Parker	71.11	9.52	60.37	12	1053_SAS_2006**
6	Barry	Bowen	Duncan	Ginobili	Parker	−25.71	8.48	−45.28	−8	1047_SAS_2006***
7	Barry	Duncan	Finley	Ginobili	Van Exel	46.74	7.70	37.35	6	4125_SAS_2006**
8	Bowen	Duncan	Finley	Ginobili	Van Exel	32.55	7.67	18.79	3	4126_SAS_2006
9	Duncan	Finley	Ginobili	Horry	Van Exel	−22.46	5.84	−24.68	−3	4156_SAS_2006
10	Barry	Bowen	Duncan	Finley	Van Exel	−53.95	5.66	−59.24	−7	4111_SAS_2006***
11	Barry	Finley	Ginobili	Horry	Parker	−14.32	5.53	−26.06	−3	1081_SAS_2006
12	Bowen	Duncan	Ginobili	Horry	Van Exel	46.51	4.13	34.84	3	4150_SAS_2006
13	Barry	Bowen	Finley	Horry	Parker	−179.83	4.11	−186.96	−16	1067_SAS_2006***

Note: *denotes the player's team; ** denotes a good lineup; *** denotes a bad lineup. Simple = Raw +/− rating per 48 minutes.

outplay Parker). During the 2006–7 season, however, Parker played better against Nash. For the season Nash had an impact of +28%, and against the Spurs Nash had an impact of 0%. For the season Parker had an impact of −3%, but against the Suns Parker had an impact of +35%. Thus, basketball matchups can exhibit a lack of transitivity. The great coaches probably intuitively understand matchups, but WINVAL analysis allows coaches to be more data driven when they make crucial decisions about how their lineups should be selected to best perform against their opponent's on-court lineup.

33

NBA PLAYERS' SALARIES AND THE DRAFT

In chapter 9 we determined salaries for baseball players based on how many wins a player generated over and above the number of wins that would be achieved with a team of "replacement players." Using the WINVAL point ratings we can use the same approach to come up with an estimate of a fair salary for an NBA player.

During the 2006–7 season the average team payroll was $66 million. The minimum player salary was around $400,000. We will define the point value of a "replacement player" as -6. This is the usual point value for a player in the bottom 10% of the league point values. A team made up of replacement players would lose by $5 \times (6) = 30$ points per game to an average NBA team. After noting that an average NBA team scored 98.7 points per game during 2006–7, we find that our team of replacement players would have a scoring ratio of $68.7/98.7 = .696$. Using the Pythagorean Theorem as we did in chapter 1, we find that our team of replacement players would be expected to win a fraction $\dfrac{.696^{13.91}}{.696^{13.91}+1} = .0064$ of their games. This means our team of replacement players would not win one game during an 82-game season.

Calculating a Fair Player Salary Based on Minutes Played and Points Rating

During an 82-game season, an NBA team plays $82 \times 48 = 3,936$ minutes (excluding overtime). To win 41 games, the team would need a scoring ratio of 1, which means that for $3,936 \times 5 = 19,680$ minutes the team would have to play 6 points better per 48 minutes of player time than a team made up of replacement players. Define Points over Replacement Player (PORP) by player point rating$-(-6)$ = player point rating of $+6$.

If the sum of all players on a team of PORP × minutes played = 19,680 × 6 = 118,080, the team should have a scoring ratio of 1 and be projected to win half their games. Since playing at a −6 point level per player requires 5 × (400,000) = $2,000,000 and generates no wins, we find that $64 million/41 = $1,560,976 (salary) equates to one win. We have just seen that if sum of PORP × minutes is 118,080, then the team wins 41 games. Thus a player with PORP × minutes equal to 118,080/41 = 2,880 generates 1 win. This implies that

$$\text{wins generated by player} = (\text{point rating} + 6) \times (\text{minutes played})/2,880.$$

Since each win is worth $1,560,976 of salary we can conclude that

$$\text{fair salary for player} = (\text{wins generated by player}) \times 1,560,976.$$

As an example, in 2006–7 Kevin Garnett had a 18.98 point rating and played 2,995 minutes. We estimate that he generated $\frac{(18.98+6)\times 2,995}{2,880} = 25.98$ wins. A fair salary for him would have been 25.98 × (1,560,976) = $40,550,000. His actual 2006–7 salary was the highest in the league ($21,000,000), but he was vastly underpaid. Figure 33.1 shows estimated and actual salaries for several other NBA players as well as Garnett. Note Vince Carter and Darko Millicic are vastly overpaid. LeBron James and Kevin Garnett are vastly underpaid, and Jason Kidd and Kobe Bryant are paid about what they are worth.

Each year NBA teams are penalized $1 for each dollar their payroll exceeds the salary cap. The tax level for the 2006–7 season was set at $65.42 million. Any team whose team salary exceeded that figure paid a $1 "luxury" tax for each $1 by which it exceeded $65.42 million. The luxury tax

	G	H	I	J	K	L
4	Player	Minutes	Points Rating	Wins	Estimated Worth	Actual Salary
5	Garnett	2995	18.98	25.97747	$40,550,189.70	$21,000,000.00
6	James	3190	15.12	19.62736	$30,637,831.98	$5,828,000.00
7	Millicic	1913	−1.02	1.049493	$1,638,233.06	$5,218,000.00
8	Bryant	3140	10.17	13.92285	$21,733,224.93	$17,718,000.00
9	Carter	3126	3.2	6.295417	$9,826,991.87	$15,101,000.00
10	Kidd	2933	7.66	10.44881	$16,310,341.46	$18,084,000.00

Figure 33.1. Estimated and actual salaries for the 2006−7 season.

makes it essential for NBA teams to manage their payroll well. Trading for players whose value exceeds what they will be paid is a good step in that direction.

Is the NBA Draft Efficient?

In chapter 27 we reviewed the Thaler-Massey (TM) study of the NFL draft. TM found that the NFL draft was inefficient in that later draft picks created more value than earlier picks. We viewed the TM results as flawed, however, because they lacked an accurate measure for player value. Winston and Medland[1] studied the efficiency of the NBA draft. For players drafted during the 1998–2002 period, we computed (using the method described earlier in the chapter) a fair salary for each of the player's first three seasons based on their point rating and minutes played. For their first three years in the league we defined the player's surplus value as (sum of player salary for first three years)–(salary paid for first three years (based on draft position). We averaged the surplus values over all draft picks by position in the draft. Table 33.1 illustrates the dependence of surplus value on

TABLE 33.1
Three-Year Surplus Value of Players as a
Function of Draft Position

Draft Position	Average Surplus (in millions)
1–5	$8.9
6–10	$9.45
11–15	$2.29
16–20	$5.23
21–25	$2.97
26–30	$3.05

Note: The 6–10 average surplus is "pumped up" by the phenomenal success of #10 draft picks Paul Pierce, Jason Terry, Joe Johnson, and Caron Butler.

[1] Leonhardt, "It's Not Where NBA Teams Draft, But Whom They Draft."

draft position. Therefore, all first-round draft positions create value, with picks 1–10 creating the most value. These data indicate that NBA draft selectors do get fair value for the players chosen.

Winston and Medland also evaluated whether college, high school, or international players were systematically over- or undervalued by draft selectors. For high school players, for example, we added up the predicted salaries for their first three years in the NBA based on the draft position (using table 33.1) of all high school players. Then we looked at the actual player surplus value of the high school players drafted and compared the results. We found that NBA teams on average overvalued high school draft picks by 31%. International players and players with two years of college experience were on average valued exactly correctly. Players with one year of college experience were on average overvalued by 28%. Players with three years of college experience were on average undervalued by 8%. Players with four years of college experience were undervalued by 13%. Therefore it appears that NBA teams overvalue younger draftees and undervalue older draftees.

ARE NBA OFFICIALS PREJUDICED?

The sports page of the May 2, 2007, edition of the *New York Times* contained the headline: "Study of NBA Sees Racial Bias in Calling Fouls."[1] The article was based on a study by Cornell professor Joseph Price and Wharton professor Justin Wolfers.[2] Price and Wolfers (PW) claim that "more personal fouls are called against players when they are officiated by an opposite-race refereeing crew than when officiated by an own-race crew." In this chapter we discuss their insightful analysis of the referee bias question.

What Are the Best Data to Use
to Test for Referee Bias?

An NBA officiating crew consists of three officials. The ideal way to determine whether the racial composition of the officiating crew influences the rate at which fouls are called against whites and blacks would be to look at a set of NBA games and determine the rate at which black officials and white officials call fouls on white and black players. The data might look something like the 1,000 games of data excerpted in figure 34.1. For each game in figure 34.1, we can classify each foul into one of four groups:

- a black official calling a foul on a black player
- a white official calling a foul on a white player
- a white official calling a foul on a black player
- a black official calling a foul on a white player

A 1 in columns E–G denotes a white official, while a 0 in columns E–G denotes a black official. For example, in game 1 there were one white and

---

[1] This article is available at http://www.nytimes.com/2007/05/02/sports/basketball/02refs
.html.
[2] Price and Wolfers, "Racial Discrimination among NBA Referees."

	C	D	E	F	G	H	I	J	K	L	M
3	Game	Whites	Ref. 1	Ref. 2	Ref. 3	Black minutes	White minutes	Black ref./ Black player	White ref./ White player	White ref./ Black player	Black ref./ White player
4	1	1	1	0	0	396.8463	83.153734	35	1	6	10
5	2	2	1	1	0	283.9803	196.01969	14	14	20	8
6	3	2	1	1	0	274.5583	205.44166	6	14	14	9
7	4	3	1	1	1	369.2381	110.76186	0	9	38	0
8	5	3	1	1	1	387.8274	92.172632	0	8	44	0
9	6	2	1	1	0	350.3648	129.63517	12	6	18	6
10	7	3	1	1	1	342.2891	137.71092	0	19	35	0
11	8	2	1	1	0	315.0947	164.90532	9	9	26	5
12	9	2	1	1	0	337.8692	142.13078	10	11	24	9

Figure 34.1. Ideal data set to test for referee bias. See file Refsim.xls.

two black officials. Blacks played around 397 minutes during the game and whites around 83 minutes. Thirty-five fouls were called by black officials against black players, 1 foul was called by the white official against a white player, 6 fouls were called by the white official against black players, and 10 fouls were called by a black official on a white player.

Combining these data over all 1,000 games we can determine the rate at which each of the four types of fouls occurs. We would find the following:

- White referees call 1.454 fouls per 48 minutes against black players, while black referees call only 1.423 fouls per 48 minutes against black players.
- Black referees call 1.708 fouls per 48 minutes against white players, while white officials call only 1.665 fouls per 48 minutes against white players.

These data show a clear (but small) racial bias of officials in terms of the calls they make. For example, 9,204 total fouls were called by black officials against black players. When a black official was on the court, black players' minutes totaled 310,413. (During a game in which black players play, for example, 200 minutes and there are, say, 2 black officials, this is counted as 400 black player–black official minutes.) Therefore, black officials call $48 \times 9,204/310,413 = 1.423$ fouls per 48 minutes on black players.

One problem with the data shown in figure 34.1 is that we need to know which official called each foul. This information is kept by the NBA and are not publicly available. The NBA says their study of this proprietary data does not show any evidence of prejudice. PW's analysis, however, tells a different story.

Price and Wolfers's Approach: Regression
Analysis with Interaction

Since PW do not know which officials made each call, they worked with box score data. For each player in each game they have a data point containing the following information: fouls per 48 minutes committed by the player; the race of the player; and the percentage of game officials who are white. PW refer to the first variable as Foul Rate, the second variable as Blackplayer, and the third variable as %Whitereferees. For example, if a black player played 32 minutes and committed 3 fouls, and there were two white officials, the data point would be $(3 \times 48/32, 1, 2/3) = (4.5, 1, 2/3)$. (PW use $1 = $ Blackplayer and $0 = $ Whiteplayer.)

Next, PW ran a regression (see chapter 3) to predict player fouls per 48 minutes from the following independent variables:

- %Whitereferees $\times$ Blackplayer
- %Whitereferees
- Blackplayer

PW weighted each data point by the number of minutes the player played in the game. They found that the following equation best fits the data.

$$\begin{aligned} \text{foul rate} = 5.10 + .182\text{Blackplayer} \times (\%\text{Whitereferees} \\ - .763\text{Blackplayer}) \\ - .204(\%\text{Whitereferees}). \end{aligned} \quad (1)$$

PW found that each of their independent variables has a p-value less than .001. This means that each independent variable is significant at the .001 level. In other words, for each of the three independent variables there is less than 1 chance in a 1,000 that the independent variable is not a useful variable for predicting Foul Rate.

So what can we conclude from the equation? Table 34.1 shows the predicted foul rate per 48 minutes for all possible values of %Whitereferees (0, 1/3, 2/3, 1) and the two possible values for Blackplayer (1 or 0). The table shows that when there are three black officials, blacks are called for 0.76 fewer fouls per 48 minutes than are whites, while when there is an all-white officiating crew, blacks are only called for 0.58 fewer fouls per 48 minutes than are whites. These data imply that for any composition of the officiating crew, whites commit more fouls per 48 minutes than blacks. The discrepancy between the rate at which blacks and whites foul shrinks by 23% (.18/.76) as the officiating crew shifts from all black to all white.

TABLE 34.1
Predicted Foul Rate per 48 Minutes

%Whitereferees	Blackplayer = 1	Blackplayer = 0	Black-White Rate
0	$5.10 + .182(0)(1) - .763(1) - .204(0) = 4.337$	$5.10 - .204(0) = 5.1$	-0.763
1/3	$5.10 + .182(1/3)(1) - .763(1) - .204(1/3) = 4.329$	$5.10 - .204(1/3) = 5.032$	-0.702
2/3	$5.10 + .182(2/3)(1) - .763(1) - .204(2/3) = 4.323$	$5.10 - .204(2/3) = 4.964$	-0.642
1	$5.10 + .182(1)(1) - .763(1) - .204(1) = 4.315$	$5.10 - .204(1) = 4.896$	-0.581

PW included other independent variables such as player height and weight; whether the player's team was in contention; whether the player is an All Star; and player position. Including these extra independent variables did not change the conclusion that the racial makeup of the officiating crew has a small (but statistically significant effect) on the frequency with which black and white players commit fouls.

The Meaning of a Significant Interaction

Essentially, PW found that when predicting foul rates there is a significant interaction between a player's race and the racial makeup of the officiating crew. Two independent variables interact when the effect on a dependent variable of changing one independent variable depends on the value of the other independent variable. In the PW analysis the predicted foul differential based on a player's race depends on the racial makeup of the officiating crew; the more whites in the crew the smaller the black-white discrepancy in foul rate.

To illustrate significant interaction in another way, consider the fact that price and advertising often interact when they are used to predict product sales. For example, when prices are low advertising may have a large effect on sales, but when prices are high advertising may have virtually no effect on sales. The way PW spotted the player race – official race interaction was by including the new independent variable Blackplayer × (%Whitereferees), which multiplied two independent variables. In general, to determine whether two independent variables interact, you include the product of the two independent variables as an independent variable in the regression. If the product term is significant (has a p-value less than .05), then the two independent variables do interact. If the product term has a p-value greater than .05, the independent variables do not exhibit a significant interaction.

ARE COLLEGE BASKETBALL GAMES FIXED?

In 2006 Wharton professor Justin Wolfers created a stir by claiming that around 5% of all college basketball games are fixed by players who intentionally slacken their effort (often called point shaving). Wolfers argued that for games in which the favorite is favored by S points, we should find that the probability of the favorite winning by between 1 and S − 1 points should equal the probability of the favorite winning by between S + 1 and 2S − 1 points. This follows because statisticians usually find that forecast errors about an unbiased prediction (like a point spread) should be symmetrically distributed, like a normal or Bell curve. For strong favorites (defined as teams favored by more than 12 points), Wolfers found that the forecast errors were not symmetrical about the point spread. He found that 46.2% of the time, strong favorites won by between 1 and S − 1 points, and 40.7% of the time, strong favorites won by between S + 1 and 2S − 1 points. Wolfers thus argues that 46.2% − 40.7% = 5.5% of the time players shaved points. This would account for more games ending under the spread than over the spread. The idea is once victory is in hand some of the favored team's players do not play at their optimum level. This causes more games to end with the favorite winning by a number of points in the range [1, S − 1] than the range [S + 1, 2S − 1]. Wolfers's conclusion seems a bit strong, since there may be other factors that might cause the asymmetry in forecast errors.[1]

Rebuttal

Heston and Bernhardt (HB) provide other explanations for the asymmetry in forecast errors Wolfers found for strong favorites.[2] HB looked at games

---

[1] See Wolfers, "Point Shaving in College Basketball."
[2] See Heston and Bernhardt, "No Foul Play."

involving strong favorites in which the spread increased from the opening line and games involving strong favorites in which the spread decreased from the opening line. If the game were fixed, you would expect more betting on the underdog. This would decrease the number of points by which the better team is favored. Therefore, we should expect a greater forecast error asymmetry in games in which the spread decreased than those in which the spread increased. HB found, however, that when the spread increased, 45.15% of the time the outcome was in interval $[1, S-1]$ and 39.54% of the time the outcome was in the interval $[S+1, 2S-1]$, a discrepancy of 5.61%. When the spread decreased, HB found 45.12% of the games ended in the interval $[1, S-1]$ and 39.54% of the games ended in the interval $[S+1, 2S-1]$, a discrepancy of 5.58%. Basically, whether the spread increased or decreased, the asymmetry that Wolfers found was equally present. HB also looked at Division I games in which there was no betting line. They used the Sagarin ratings (discussed in chapter 30) to set point spreads for these games and found that 48.55% of the favorites won by a margin in $[1, S-1]$ and 42.64% of the favorites won by a margin in $[S+1, 2S-1]$. The discrepancy here of 5.91% is virtually identical to the discrepancy found by HB in games in which the line increased or decreased. Since the asymmetry of outcomes about the point spread that Wolfers found persists in games where there is no gambling, it seems unlikely that "fixing the game" is the cause of this asymmetry.

HB conclude that the pervasive asymmetry in forecast errors must be a feature inherent in the way basketball is played. They provide two persuasive explanations for the asymmetry in forecast errors.

- Consider a 14-point favorite leading by 12 points late in the game. The favorite will probably start holding the ball. This will reduce the favorite's chance of covering the spread. By reducing the number of possessions in the game, holding the ball will also reduce the variability of the favorite's final victory margin. This will also reduce the favorite's chance of covering the spread.
- Key players often foul out. If a key player fouls out, then the better team is more likely to have a good replacement. This can introduce an asymmetry in the final score in terms of the predicted point spread.

DID TIM DONAGHY FIX NBA GAMES?

In July 2007 shockwaves rippled through the sports world when NBA referee Tim Donaghy was accused of fixing the outcome of NBA games. If bettors attempt to fix a game, after the opening betting line is posted the line would move substantially as the "fixers" put their bets down. A move of two or more points in the line is generally considered highly unusual. For example, on November 14, 2007, when the Toronto Raptors played the Golden State Warriors, the opening line on total points (referred to hereafter as the Total Line) was 208 points. This means that if you bet the "Over" you would win if the total points scored were 209 or more. If you bet the "Under" you would win if the total points scored were 207 points or fewer. If the total points scored were 208 neither would win. The closing Total Line was 214 points, a six-point increase. This means that near game time a large amount of money was bet on the Over. If, as was alleged, Donaghy had been trying to fix the game (causing the game to exceed the Total Line), he probably would have to have called lots of fouls. (Fouls create free throws and in effect make the game longer.) Using publicly available data, we can determine whether significantly many more free throws were attempted in games Donaghy officiated and in which the Total Line moved up two or more points than were attempted in the rest of the games Donaghy officiated. We adjusted the number of free throws attempted in each game based on the teams playing and on the other two game officials. Figure 36.1 displays the results of these calculations. For example, in the first listed game, Chicago had 32 free throw attempts and Miami had 22. An average NBA team had 26.07 free throw attempts per game during the 2006–7 season. Figure 36.2 gives the average number of free throw attempts for each team and their opponents.

The Bulls shot $(25.40 - 26.07)$ more free throws than average while the Heat shot $(24.62 - 26.07)$ fewer free throws than average. The Bulls'

	AS	AT	AU	AV	AW	AX	AY	AZ	BA	BB	BC	BD	BE
7	Away	Home	FTs (away)	FTs (home)	FTs predicted (away)	FTs predicted (home)	Ref. effect	Total FTs predicted (away)	Total FTs predicted (home)	Total predicted FTs	Away Delta	Home Delta	Total Delta
8	Chicago	Miami	32	22	27.86902439	23.564146	−0.17	27.784024	23.479146	51.26317	4.216	−1.48	2.7368293
9	Dallas	Houston	26	23	18.58853659	22.064146	2.98	20.078537	23.554146	43.63268	5.921	−0.55	5.3673171
10	Dallas	Portland	25	25	19.80804878	24.661707	−1.48	19.068049	23.921707	42.98976	5.932	1.078	7.0102439
11	Toronto	Golden State	28	23	25.8202439	21.832439	−1.48	25.080244	21.092439	46.17268	2.92	1.908	4.8273171
12	Phoenix	Utah	32	33	31.01536585	20.600732	−0.67	30.680366	20.265732	61.52463	4.158	9.317	13.475366
13	Chicago	New York	37	38	32.46658537	28.308049	0.75	32.841585	28.683049	55.67317	3.267	1.06	4.3268293
14	Memphis	L.A. Clippers	27	33	24.61292683	32.820244	−1.76	23.732927	31.940244	55.67317	3.267	1.06	4.3268293
15	Orlando	Portland	20	21	24.44219512	28.90561	−0.75	24.067195	28.53061	52.5978	−4.07	−7.53	−11.5978
16	Philadelphia	Chicago	26	24	25.63731707	29.283659	0.92	26.097317	29.743659	55.84098	−0.1	−5.74	−5.840976
17	Utah	Minnesota	22	24	21.39341463	32.247073	−0.5	21.143415	31.997073	53.14049	0.857	−8	−7.140488
18	Detroit	New Jersey	23	27	24.6495122	21.661707	−3.3	22.999512	20.011707	43.01122	5E−04	6.988	6.9887805
19	Washington	Denver	18	44	29.35682927	28.369024	1.66	30.186829	29.199024	59.38585	−12.2	14.8	2.6141463
20	Dallas	Seattle	21	30	18.6495122	27.051951	−0.5	18.399512	26.801951	45.20146	2.6	3.198	5.7985366
21	L.A. Clippers	Houston	30	26	26.67390244	24.503171	1.91	27.628902	25.458171	53.08707	2.371	0.542	2.9129268
22	Memphis	Washington	37	24	26.80804878	28.88122	1.91	27.763049	29.83622	57.59927	9.237	−5.84	3.4007317

Figure 36.1. Free throws attempted with Donaghy officiating. See file Donaghy.xls.

	V	W	X
6	**Team**	**By**	**Against**
7	Phoenix	22.3	21.45122
8	Dallas	24.9	22.08537
9	Utah	30	24.5122
10	Washington	29.6	23.5
11	San Antonio	24.1	21.20732
12	Detroit	24.2	20.34146
13	L.A. Lakers	27.1	23.36585
14	Denver	29.9	23.73171
15	Toronto	24.2	21.78049
16	Seattle	23.3	19.93902
17	Golden State	25.8	22.34146
18	Memphis	29.4	24.29268
19	Milwaukee	23.4	20.95122
20	Houston	23.2	19.89024
21	Kansas City	30.2	24.26829
22	New Jersey	26.3	23
23	New York	29.2	22.86585
24	Cleveland	26	22
25	L.A. Clippers	27.4	22.30488
26	Chicago	25.4	23.17073
27	Miami	24.6	21.82927
28	Portland	24.5	21.34146
29	New Orleans	23.7	21.13415
30	Orlando	29.2	23.42683
31	Minnesota	23.1	20.47561
32	Philadelphia	26	21.96341
33	Charlotte	26	21.78049
34	Atlanta	26.9	22.59756
35	Boston	26.6	21.42683
36	Indiana	25.9	23.68293

Figure 36.2. Free throws attempted, 2006−7 season, by team.

opponents shot (23.17 − 26.07) more free throws than average while the Heat's opponents shot (21.83 − 26.07) fewer free throws than average. Using this data we would predict the Bulls to have made 26.07 + (25.40 − 26.07) + (21.83 − 26.07) = 25.40 + 21.83 − 26.07 = 21.16 free throw attempts. Similarly, we estimated the Heat would attempt 21.72 free throws. Based on 2006–7 NBA game data, we estimated the other two officials in this game combined on average (in their other games) to have allowed 0.20 fewer free throws than average. Therefore, our predicted total number of free throw attempts for the game would be 21.16 + 21.72 − 0.2 = 42.68. Since 54 free throws were actually attempted, in this game the teams attempted 54 − 42.68 = 11.32 more free throws than expected. We call actual free throw attempts in a game minus the predicted free throw attempts in a game delta free throws. Over the eleven games Donaghy officiated in which the Total Line moved up two points or more, 180.27 more free throws than expected were attempted (an average of 16.39 more free throws per game). In Donaghy's remaining games, 460.86 more free throws than expected were attempted (an average of 7.32 more free throws per game). If Donaghy's propensity for calling fouls was the same in games in which the Total Line moved at least two points as during his remaining games, what is the probability that a discrepancy at least this large would have occurred?

The standard deviation of the average of delta free throw attempts over n games is $11.39/n^{.5}$ where 11.39 is the standard deviation of delta free throw attempts over all games. Then the standard deviation of (mean delta free throw attempts in games in which the Total Line moved two or more points)−(mean delta free throw attempts in games in which the Total Line did not move two or more points) is given by $\sqrt{\dfrac{11.39^2}{11} + \dfrac{11.39^2}{63}} = 3.72$. The observed value of (mean delta free throw attempts in games in which the Total Line moved two or more points)−(mean delta free throw attempts in games in which the Total Line did not move two or more points) points was 16.39 − 7.32 = 9.07. The probability of a discrepancy in mean free throw attempts of 9.07 or larger is simply the probability that a normal random variable with mean 0 and standard deviation 3.72 exceeds 9.07. This probability can be computed in Excel as 1 − NORMDIST (9.07,0,3.72,True) = .005. This means that if the mean free throw attempts in games Donaghy officiated and the spread moved two or more points was identical to the mean free throw attempts in his other games, a discrepancy in free throw attempts at least as large as what we observed would only occur with a probability of 1/200. This analysis conclusively indicates that in games Donaghy officiated and the Total Line increased by at least two points, significantly more free throws were attempted than were attempted in other games in which Donaghy officiated.

A Better Analysis

The NBA knows exactly how many fouls each official calls during each game. (They do not release these data.) A better test of whether Donaghy fixed games would be to compare the percentage of fouls Donaghy called in games where the Total Line increased by at least two points to the percentage of fouls he called in all other games. If this difference is statistically significant, this would be additional evidence to support the hypothesis that during games in which the Total Line increased by at least two points, Donaghy called more fouls in an effort to cause the game total points to exceed the Las Vegas Total Line.

END-GAME BASKETBALL STRATEGY

In this chapter we will consider the optimal strategy for two important situations that can occur at the end of a close basketball game:

- In game 1 of the first round of the 2001 Eastern Conference playoffs the Philadelphia 76ers led the Indiana Pacers by two points. The Pacers had the ball with five seconds to go. Should the Pacers have attempted a two-pointer to tie or a three-pointer to win?
- During game 6 of the 2005 Western Conference semifinals, the Dallas Mavericks led the Phoenix Suns by three points with five seconds to go. Steve Nash is bringing the ball up the court. Should the Mavericks foul Nash or allow him to attempt a game-tying three-pointer?

I was fortunate enough to be at both of these exciting games. Reggie Miller hit a game-winning three-pointer as the buzzer went off in the Pacers–76ers game. Steve Nash hit a game-tying three-pointer and the Suns went on to eliminate the Mavericks in a double overtime thriller.

Let's use mathematics to analyze the optimal strategy in both of these exciting end-of-game situations.

Should A Team Go for Two or Three Points?

To begin, let's assume we have the ball and we trail by two points with little time remaining in the game. Should our primary goal be to attempt a game-tying two-pointer or to go for a buzzer-beating three-pointer to win the game? This situation has often been used in Microsoft job interviews.[1] We assume our goal is to maximize the probability that we win the game. To simplify matters, we assume that no foul will occur on our shot and that

---

[1] Thanks to Norm Tonina for sharing this fact with me.

the game will end with our shot. To make the proper decision, we will need to estimate the values of the following parameters:

- PTWO = Probability that a two-pointer is good. For the entire season PTWO is around .45.
- PTHREE = Probability a three-pointer is good. For the entire season PTHREE is around .33.
- POT = Probability that we will win the game if the game goes to overtime. It seems reasonable to estimate POT to be near .5.

If we go for two points we will win the game only if we hit the two-pointer and win in overtime. Events are said to be independent if knowing that one event occurs tells us nothing about the probability of the other event occurring. To find the probability that independent events $E_1, E_2, \ldots E_n$ will all occur, we simply multiply their probabilities. Since hitting a two-pointer and then winning in overtime are independent events, the probability of winning the game if we go for two points is PTWO $\times$ POT = $(.45)(.5) = .225$.

If we go for three points, we win if and only if we make the three-pointer. This will happen with probability PTHREE = .33. Thus it seems that we have a much better chance of winning the game if we go for three. Most coaches think this is the riskier strategy and go for two points.

Of course, our parameter estimates may be incorrect, so we should perform a sensitivity analysis to determine how much our parameters would have to change for our optimal decision to change. A standard way of performing a sensitivity analysis is to hold all but one parameter constant at its most likely value and determine the range on the remaining parameter for which our optimal decision (going for three) remains optimal.

We find that as long as PTWO is less than .66, we should go for three if PTHREE = .33 and POT = .5. We find that as long as PTHREE is greater than .225, we should go for three if PTWO = .45 and POT = .5. We find that if POT is less than $33/45 = .733$, we should go for three if PTWO = .45 and PTHREE = .33.

This sensitivity analysis makes it clear that our parameter estimates would have to be greatly in error to make going for two points the optimal strategy.

Should We Foul When We Lead by Three Points?

Let's address whether a team with a three-point lead should foul the other team when little time remains on the clock. This is a much tougher question (and I think the definitive analysis remains to be done).

Two sports analysts, Adrian Lawhorn and David Annis, have analyzed this situation and both have concluded that the defensive team should foul the team with the ball.[2] Let's briefly discuss their logic.

Lawhorn begins by looking at actual data. He assumes that the current possession will be the last possession during regulation. He finds that with less than 11 seconds left, teams that trailed by three points hit 41 out of 205 three-pointers. This means that if the team with the ball is allowed to shoot, they have a 20% chance of tying the game and therefore a 10% chance of winning the game (assuming each team has the same chance to win in overtime). If the defensive team fouls the trailing team, Lawhorn assumes that the only way the trailing team can tie is to make the first free throw and then miss the second free throw intentionally. Then Lawhorn assumes the trailing team must either tip the ball in or kick the ball out and hit a two-pointer to tie the game. Lawhorn calculates that the trailing team has roughly a 5% chance of tying the game or around a 2.5% chance of winning the game. Lawhorn does not consider the ultimate coach's nightmare: the trailing team hits the first free throw and after rebounding an intentionally missed second shot hits a three-pointer to win. NBA teams hit around 75% of their free throws and rebound around 14% of missed free throws. Suppose the trailing team can hit 30% of their three-pointers. If the trailing team elects to take a three-pointer after rebounding a missed three-pointer, then the probability that they will win is $.75(.14)(.3) = .03$, or 3%. This is slightly better than the 2.5% chance of winning if they attempt a two-pointer. Therefore, fouling the trailing team reduces their chances of winning from 10% to 3% and seems like a good idea. Annis comes to a similar conclusion. The key assumption made by Lawhorn and Annis is, of course, that the game ends after the current possession. In reality, if the defensive team fouls the trailing team after their free throws, they in turn will foul the defensive team and probably get the ball back. If the trailing team makes both free throws and the defensive team misses one out of two free throws, then the trailing team can win the game with a

---

[2] See Lawhorn, " '3-D': Late-Game Defensive Strategy with a 3-Point Lead," and Annis, "Optimal End-Game Strategy in Basketball."

three-pointer or tie with a two-pointer! The problem is that we cannot know whether the current possession is the last possession. Lawhorn found 32 games in which a team trailed by three points with less than 11 seconds left and the leading team intentionally fouled the trailing team. In seven of these games (21.9%), the trailing team tied the game. Although this is a small sample, the data indicate that when multiple possessions are a possibility, fouling may not be the correct strategy.

A student in my sports and math class, Kevin Klocke, looked at all NBA games from 2005 through 2008 in which a team had the ball with 1–10 seconds left and trailed by three points.[3] The leading team did not foul 260 times and won 91.9% of the games. The leading team did foul 27 times and won 88.9% of the games. This seems to indicate that fouling does not significantly increase a team's chance of winning when they are three points ahead.[4]

[3] Kevin Klocke, "Basketball 3-Point Strategy," unpublished paper, Indiana University.

[4] We believe more work needs to be done to determine the definitive answer to this question. We are working on a simulation model of the last minute of a basketball game that should help settle the issue.

PART IV

PLAYING WITH MONEY, AND OTHER TOPICS FOR SERIOUS SPORTS FANS

SPORTS GAMBLING 101

In this chapter we will review (largely through a question-and-answer format) the basic definitions and concepts involved in football, basketball, and baseball gambling.

Betting on the Odds

In the 2007 Super Bowl the Colts were favored by 7 points over the Bears and the predicted total points for the game was 48. How could someone have bet on these odds? Theoretically the fact that the Colts were favored by 7 points meant the bookies thought there was an equal chance that the Colts would win by more than 7 or less than 7 points (in the next chapter we will see this may not be the case). We often express this line as Colts -7 or Bears $+7$, because if the Colts' -7 points is greater than 0, a Colts' bettor wins, while if the Bears' $+7$ is greater than 0, a Bears' bettor wins.

Most bookmakers give 11–10 odds. This means that if we bet "a unit" on the Colts to cover the point spread, then we win $10 if the Colts win by more than 7 points. If the Colts win by fewer than 7 points we pay the bookmaker $11. If the Colts win by exactly 7 points, the game is considered a "push" and no money changes hands. A total points bet works in a similar fashion. If we bet the "over" on a totals bet we win $10 if more than 48 points are scored, while we lose $11 if the total points scored is fewer than 48. If exactly 48 points are scored the totals is a push and no money changes hands. Similarly, if a bettor takes the under he wins if the total points scored are fewer than 48 and he loses if the total points scored are greater than 48. Most gamblers believe the Total Line (in this case, 48 points) is the most likely value of the total points scored in the game. Basketball point spread betting and totals betting work in an identical fashion to football betting.

How Do Bettors Make Money Gambling?

Let p = probability that a gambler wins a point spread bet. If $10p - 11(1 - p) = 0$, our expected profit on a bet equals 0. We find that $p = 11/21 = .524$ makes our expected profit per bet equal to 0. Therefore, if we can beat the spread or totals more than 52.4% of the time we can make money. Suppose we are really good at picking games and can win 57% of our bets. What would be our expected profit per dollar invested? Our expected return per dollar invested is $(.57(10) + .43(-11))/11 = 8.8\%$. Thus if we can pick winners 57% of the time we can make a pretty good living betting. However, doing this against the spread in the long run is virtually impossible. If we believe we have a probability $p > .524$ of winning a bet, what percentage of our bankroll should we bet on each gamble? In chapter 44 we will use the famous Kelly growth criteria to answer this question.

How Do Bookmakers Make Money?

Until Steven Levitt's brilliant article on NFL betting (discussed in chapter 39) was published in 2004, the prevailing wisdom was that bookmakers tried to set the line so half the money was bet on each side.[1] If this is the case, the bookmaker cannot lose! For example, suppose one person bets $10 on the Colts −7 (Colts to win by 7 points) and one person bets $10 on the Bears +7 (this means if the Bears' +7 points is larger than the Colts' points, the underdog bettor wins). Then unless the game is a push, the bookie pays one bettor $10 and collects $11 from the other bettor and is guaranteed a profit of $1. The bookmaker's mean profit per dollar bet is called vigorish or "the vig." In our example, $11 + 11 = \$22$ is bet, and the bookmaker wins $1 so the vig is $1/22 = 4.5\%$. In our example the bookmaker makes a riskless profit of 4.5%. We will see in the next chapter that a smart bookmaker can take advantage of gambler biases and make an expected profit (with some risk, however) that exceeds 4.5%.

How Does the Money Line Work?

The money line enables a bettor to bet on who wins a game or an event, not the margin of victory. For example, the money line on the 2007 NBA Finals was Spurs −450, Cavaliers +325. For any money line the team with

[1] Levitt, "Why Are Gambling Markets Organised So Differently from Financial Markets?"

the negative number is the favorite and the team with the positive number is the underdog. The meaning of this money line is that to win $100 on the Spurs you must bet $450. Thus if the Spurs win the series I win $100 but if the Spurs lose I lose $450. If you bet $100 on the Cavaliers you win $325 if the Cavs win and lose $100 if the Cavs lose. Let p = probability that the Spurs win the series. A gambler who believes that $100p - 450(1 - p) > 0$ would bet on the Spurs while a gambler who believes that $325(1 - p) - 100p > 0$ would bet on the Cavaliers. Solving for the value of p that satisfies each inequality, we find that gamblers who feel the Spurs have a chance greater than $9/11 = 82\%$ of winning would bet on the Spurs, while gamblers who feel that the Spurs have a $p < 13/17 = 76\%$ chance of winning would bet on the Cavs. If we assume that the true probability of the Spurs winning was the average of 76% and 82% (79%) and also assume that bettor estimates of the Spurs' chances of winning are symmetrically distributed about 79%, then we would expect an equal number of bettors to bet on Cleveland and San Antonio. Suppose one gambler bets on the Spurs and one on the Cavs. If the Spurs win the bookmaker breaks even by paying the Spurs bettor $100 and collects $100 from the Cavs bettor. If the Cavs win the bookmaker wins $125 by collecting $450 from the Spurs bettor and paying $325 to the Cavs bettor. If the Spurs' true chance of winning the series is 79%, then the gambler's expected profit per dollar bet is given by $(.79(0) + .21(125))/(450 + 100) = 4.8\%$. We will learn in chapter 40 how to use point spreads to estimate probabilities of a team winning a game, an NBA playoff series, or the NCAA tournament.

The 2007 Super Bowl provides another example of how the money line works. For this game the money line with one Internet bookmaker was Colts −250 and Bears +200. That means if you bet the Colts to win and they lose you lose $250, and if the Colts win you win $100. If you bet the Bears to win and they win you win $200, but if the Bears lose you lose $100.

How Does Baseball Betting Work?

The starting pitchers play a critical role in determining the winner of a game. Therefore, the baseball gambling line is only valid as long as the listed starting pitchers actually start the game. A sample baseball betting line might look like the one illustrated in table 38.1. In this example, the Yankees are playing the Red Sox at Fenway Park with Mike Mussina starting for the

TABLE 38.1
Sample Baseball Betting Line

Date and Time	Teams (Away First Then Home)	Starting Pitchers	Money Line	Total
July 12, 1 P.M.	Yankees	Mike Mussina	$+135$	over 7.5, -115
	Red Sox	Josh Beckett	-145	under 7.5, $+105$

Yankees and Josh Beckett for the Red Sox. Again the team with the negative entry in the money line is the favorite and the team with the positive entry is the underdog. If you bet $145 on the Red Sox and they win you win $100, while if you bet $100 on the Yankees to win and they win you win $135. Following the logic in our Spurs-Cavs example, the interested reader can show that a gambler would bet on the Red Sox if she believes the Red Sox's chance of winning exceeds 29/49 = 59%, and a gambler would bet on the Yankees if she believes the Red Sox's chance of winning is less than 135/235 = 57%.

The Total column is analogous to the Total Line in football or basketball. Over 7.5 runs plays the role of favorite. If we bet $115 on the Over we win $100 if 8 or more total runs are scored in the game. If 7 or fewer runs are scored, then we lose $115. If we bet the under side of the bet we lose $100 if 8 or more runs are scored. If 7 or fewer runs are scored we win $105.

What Is an Arbitrage Betting Opportunity?

Often different bookmakers and Internet betting sites have lines on games that differ slightly. In rare cases, a combination of bets exists (called an arbitrage opportunity) that guarantees you a riskless profit. For example, suppose two different bookies had the following lines on the 2007 Super Bowl:

Bookie 1	Colts -122	Bears $+112$
Bookie 2	Colts -135	Bears $+125$

Since Bookie 1 offers better odds on the Colts and Bookie 2 offers better odds on the Bears, we will bet on the Colts with Bookie 1 and the Bears with Bookie 2. Suppose we bet x with Bookie 1 on the Colts and suppose you bet $100 with Bookie 2 on the Bears. If the Colts win your profit is $100 \times (x/122) - 100$. This will be greater than 0 if x > $122. If the Bears win your profit is $125 - x$, which is greater than 0 if $x < 125$. Therefore, by betting $100 on the Bears and between $122 and $125 on the Colts, we can lock in a sure profit. For example, betting $123.50 on the Colts and $100 on the Bears locks in a sure profit of $1.23. The problem with an arbitrage opportunity is that the line can move before you finish placing all the needed bets. For example, if after betting $100 on the Bears with Bookie 2 the Colts line with Bookie 1 moves to -130 before we can place our bet, then an arbitrage opportunity no longer exists.

What Is a Parlay?

A parlay is a selection of two or more bets, all of which must win for the parlay to pay off. If any of the bets result in a push, no money changes hands. An example of a two-bet parlay would be taking the Colts -4 to beat the Patriots and the Bears -6 to beat the Saints. You can combine total bets with point spreads and even combine bets involving different sports.

The true odds and the typical payout on parlays are shown in table 38.2. For example, in a two-team parlay we have a 1/2 chance of winning each bet so our chance of winning the parlay (ignoring a push on either bet) is $(1/2)^2 = .25$. Three-to-one odds would be fair because then our expected profit on a $100 bet would be $.25(300) - .75(100) = \$0$. With an actual payout of 2.6–1 our expected profit on a $100 bet is $.25(260) - .75(100) = -\$10$, which is an average house edge of -10%. The more teams in the parlay, the larger the house edge.

Our calculation of the house edge assumes that the bets are independent, that is, the outcome of one bet does not affect the outcome of the other bet.

TABLE 38.2
Parlay Betting Payoffs

Number of Bets	Actual Odds	Standard Payout Odds	House Percentage Edge
2	3–1	2.6–1	10
3	7–1	6–1	12.50
4	15–1	12–1	18.75
5	31–1	25–1	18.75
6	63–1	35–1	43.75

For example, the results of bets on the point spreads of two different games would be independent. If we were to choose a two-bet parlay involving the Colts −7 points over the Titans and the total points over on a line of 44 points, these bets might not be independent. Our logic might be that if the Colts cover the point spread, then Peyton Manning must have had a good day and the total points is more likely to go over 44. Looking at it another way, if the Colts fail to cover, it was probably a bad day for Manning and our over bet has little chance of winning. This is an example of a correlated parlay, because the outcomes of the bets composing the parlay are correlated. Suppose that if the Colts cover there is a 70% chance the total will go over 44 points, while if the Colts do not cover there is only a 30% chance that the total will go over 44 points. Then our chance of winning the parlay is .5(.7) = .35, which is far better than our chance of winning a two-bet parlay composed of independent bets. For this reason, most bookmakers will not take correlated parlays.

What Are Teasers?

To illustrate a teaser bet, suppose that in one game the San Diego Chargers are an eight-point favorite and in another game the Tennessee Titans are a three-point underdog. A two-team 7-point teaser involving these games makes your point spread 7 points "better" on each game but you need to win both bets to collect. Thus if we place a 7-point teaser bet on these games and take the Chargers and Titans, we win if and only if the Chargers win by more than one point and the Titans lose by nine or fewer points. If either game ends with a tie against the revised point spread the teaser is called off and no money changes hands. Otherwise we lose the teaser bet. Here are some examples of how this teaser might play out:

- Chargers win by two and Titans lose by three—we win the teaser.
- Chargers win by three and Titans lose by twelve—we lose the teaser.
- Chargers win by one—the teaser is a push and no money changes hands.
- Chargers lose by one and Titans win by five—we lose the teaser.

Teasers usually involve 6, 6.5, or 7 points. Table 38.3 shows some typical teaser payoffs. For example, if we bet a two-team 7-point teaser and we win the teaser, then we win $100. If we lose the teaser we lose $130. If we bet a four-team 6-point teaser and all four teams cover their revised points, we win $300. If no games push and we do not cover all four revised spreads we lose $100. During the years 2000–2005 teams covered 7-point

TABLE 38.3
Teaser Payoffs

Number of Teams	6-Point Teaser	6.5-Point Teaser	7-Point Teaser
2	−110	−120	−130
3	+180	+160	+150
4	+300	+250	+200
5	+450	+400	+350
6	+700	+600	+500

Source: Stanford Wong, Sharp Sports Betting (Pi Yee Press, 2001).

teasers 70.6% of the time, pushed 1.5% of the time, and lost 27.9% of the time.

Let's determine our expected profit on a two-team teaser with a $100 bet. We begin by figuring the probability that we win the teaser, push, or lose the teaser. We assume that the outcomes of the individual teaser bets are independent events. That is, if we cover one game involved in the teaser, this does not affect our probability of covering any other game involved in the teaser. Now we can compute the probability of winning a two-team teaser bet.

- We win the teaser with probability $.706^2 = .498436$.
- We push if exactly one game is a push or both games push. This occurs with probability $(.015) \times (1 − .985) + (.985) \times (.015) + (.015)^2 = .043516$.
- We lose the teaser with probability $1 − .498436 − .043516 = 0.458048$.

Our expected profit on the teaser is $(100)(.498436) + 0(.043516) − 130(.458048) = −9.7$.

Therefore a two-team teaser is favorable to the bookmaker and unfavorable to the bettor. A little algebra shows that (assuming the push probability remains the same) we need the probability of covering a game involved in the teaser to increase by 3% (to 73.6%) in order to break even. The interested reader can show that the bookie's edge increases as more teams are involved in the teaser.

39

FREAKONOMICS MEETS THE BOOKMAKER

Recall from chapter 38 that if a bookmaker gives 11–10 odds on NFL point spread bets and sets a line so that half the money is bet on each side, then the bookmaker is guaranteed to make a riskless 4.5% profit.

Steven Levitt of *Freakonomics* fame showed that bookmakers can exploit bettor biases to make an expected profit exceeding 4.5% per dollar bet. Levitt obtained bettor records for 20,000 bettors during the 2001 NFL season. He found that much more than 50% of all money is bet on favorites and less than 50% on underdogs. When the home team was favored, 56.1% of the bets were on the favorite and 43.9% of the bets were on the underdog. When the visiting team was favored, 68.2% of the bets were on the favorite and 31.8% of the bets were on the underdog. These results are inconsistent with the widely held belief that bookmakers set spreads in an attempt to balance the amount of money bet on the underdog and favorite. We will see that if more money is bet on favorites, and favorites cover the spread less than half the time, then bookmakers can earn an expected profit exceeding 4.5%. For Levitt's sample, this does turn out to be the case. In Levitt's sample, bets on home favorites win 49.1% of time. Bets on home underdogs win 57.7% of time. Levitt found that bets on visiting favorites win 47.8% of time, while bets on road underdogs win 50.4% of time. Thus favorites are not a good bet. These data indicate that the line on favorites is inflated to take advantage of the bias of bettors toward favorites. For example, when setting the line for Super Bowl XLI the bookmaker may have thought the Colts were only 6 points better than the Bears. Since bettors are biased toward the favorite, the bookmaker might set the line at Colts −7. Since the true situation is that the Colts are 6 points better than the Bears, a bet on Colts −7 has less than a 50% chance of winning. Because bettors are biased toward favorites, in all likelihood lots of money will still come in on the Colts. Since the true point spread should be 6, the Colts have less than a 50% chance of covering the 7-point spread. This means that on average the bookies will do better than the sure profit rate of 4.5% that they would be guaranteed if an equal amount of money were bet on each side of the line.

So how does the bookmaker do in Levitt's sample? We find that the bettors win 49.45% of their bets and lose 50.55% of their bets. Thus on average the bookmaker earns $.4945(-10) + .5055(11) = 61.56$ cents per $10 bet. If bettors win half their bets, then bookmakers make on average $.5(-10) + .5(11) = 50$ cents per $10 bet. Thus the bookies' apparent slight edge (49.45% vs. 50% wins) in winning bets translates to a $61.56/50 = 23\%$ increase in mean profits.

Levitt checked to see whether the surprising failure of favorites to cover the spread had occurred during previous seasons. During 1980–2001, 48.8% of home favorites and 46.7% of visiting favorites covered the spread. This means that a bettor could have made money by simply betting on home underdogs. Levitt also found that the tendency of the favorite to fail to cover the spread was virtually independent of the size of the spread. During 1980–2001, 48.5% of teams favored by more than 6 points covered, 48.1% of teams favored by 3.5–6 points covered, and 47.8% of teams favored by less than 3 points covered the spread.

Beating the NBA Total Points Line

In chapter 36 we saw how Tim Donaghy manipulated the final total points in NBA games by calling lots of fouls. Officials who call lots of fouls create higher-scoring games for two reasons. First, more free throws are shot, and free throws result in points. Second, when more fouls are called players cannot play defense as tightly, and this also leads to higher-scoring games. The Web site Covers.com allows us to see how each NBA official performed against the Total Line for seasons 2003–4 through 2007–8. These data are summarized in figure 39.1.

For example, when Jim Clark officiated, teams went over the Total Line 221 times and under 155 times. Thus a bet on the "over" against the Total Line when Clark officiated would have won 58.8% of the time. Is this statistically significant? Recall that an outcome that deviates from the expected outcome by two or more points indicates statistical significance. Here we have a binomial experiment, which consists of repeated independent trials in which each trial results in a success or failure and the probability of success is the same on each trial. In our example,

	D	E	F	G	H	I	J	K
6	Name	Over	Under	Total	Percentage over Bet	Mean	Sigma	z Score
7	Jim Clark	221	155	376	0.587765957	188	9.69535971	3.40369011
8	Pat Fraher	177	131	308	0.574675325	154	8.77496439	2.62109326
9	Derrick Stafford	190	149	339	0.560471976	169.5	9.20597632	2.22681433
10	Pat Joe Forte	209	167	376	0.555851064	188	9.69535971	2.16598462
11	Tommy Nunez	127	100	227	0.559471366	113.5	7.53325959	1.79205294
12	Olandis Poole	126	100	226	0.557522124	113	7.51664819	1.72949427
13	Joe DeRosa	191	162	353	0.541076487	176.5	9.39414711	1.54351426
14	Phil Robinson	166	141	307	0.540716612	153.5	8.76070773	1.42682536
15	Jess Kersey	140	120	260	0.538461538	130	8.06225775	1.24034735
16	Greg Willard	191	175	366	0.521857923	183	9.56556323	0.8363334
17	Derrick Collins	155	141	296	0.523648649	148	8.60232527	0.81373347
18	Bernie Fryer	133	122	255	0.521568627	127.5	7.98435971	0.68884672
50	David Jones	153	177	330	0.463636364	165	9.08295106	−1.3211565
51	Steve Javie	175	204	379	0.461741425	189.5	9.73396117	−1.4896299
52	Bennie Adams	149	176	325	0.458461538	162.5	9.01387819	−1.4976905
53	Eddie F.	148	175	323	0.458204334	161.5	8.98610038	−1.5023202
54	Ed Malloy	144	171	315	0.457142857	157.5	8.87411967	−1.5212777
55	Ron Garretson	176	208	384	0.458333333	192	9.79795897	−1.6329932
56	Bob Delaney	169	201	370	0.456756757	185	9.61769203	−1.6636008
57	Jack Nies	157	191	348	0.451149425	174	9.32737905	−1.8225913
58	Mike Callahan	166	218	384	0.432291667	192	9.79795897	−2.6536139
59	Derek Richardson	143	195	338	0.423076923	169	9.19238816	−2.8284271
60	Kevin Fehr	127	181	308	0.412337662	154	8.77496439	−3.0769356
61	Ron Olesiak	136	203	339	0.401179941	169.5	9.20597632	−3.6389405

Figure 39.1. Performance data for NBA officials, 2003–8. Source: Data from Covers.com.

- a trial is a game officiated by the referee in question;
- a success occurs when a game score goes over the Total Line and a failure occurs when the game score goes under the Total Line; and
- if the Total Line is incorporating all available information, then the probability of success on each trial is .50.

In a binomial experiment with n trials and probability of success p on each trial, the number of successes is closely approximated (when $np(1-p) \geq 10$) by a normal random variable with mean np and variance $np(1-p)$. Therefore, for any official we can compute a z score (see chapter 11) that indicates the number of standard deviations above or below average corresponding to each official's performance against the Total Line. For example, for Jim Clark, the mean = $.5(376) = 188$, and the standard deviation = $SQRT(376 \times .5 \times (1 - .5)) = 9.70$. Thus Jim Clark's z score is $(221 - 188)/9.70 = 3.40$. This means that if the Total Line is fair, covering the Total Line 221 times or more is equivalent to seeing an observation from a normal random variable that is 3.4 standard deviations above average. Putting this in perspective, this is about as rare as seeing a randomly chosen person whose IQ exceeds $100 + 3.4(15) = 151$ (IQs have a mean of 100 and a standard deviation of 15).

Our data show, for example, that when Jim Clark officiates games are significantly higher scoring than expected, while when Ron Olesiak is officiating, games are significantly lower scoring than expected.

These data seem to indicate that certain officials can influence the total points scored in a game. Can we make money exploiting the fact that the Total Line does not appear to adjust for this fact? Let's try the following simple system. For the 2006–7 season we will predict the probability that the over bet will win in a given game by averaging the fraction of the time (for the 2003–4, 2004–5, and 2005–6 seasons) that the over bet covered for each game official. For example, referees 1, 2, and 3 are officiating a game. When referee 1 officiated the over bet covered 45% of the time, when referee 2 officiated the over bet won 47% of the time, and when referee 3 officiated the over bet covered 49% of the time. Averaging these three percentages yields an estimate that the over bet will win 49% of the time. Let's assume that when our prediction exceeds 53% we will bet the over on the Total Line, and when our prediction is under 47% we will bet the under on the Total Line. During the 2006–7 seasons this strategy won 97 of 185 (or 52.4%) of all bets on the Total Line. Thus, assuming that a referee's past tendencies on the Total Line will continue allows us to win more than half of our bets but not to make money.

40

RATING SPORTS TEAMS

Most gamblers believe that when bookies set point spreads their goal is to have half the money bet on each team. If I bet $10, for example, on a 7.5-point favorite to cover the spread, I win $10 if the team covers but I lose $11 if the favorite does not cover. If the favorite covers the points spread half the time, on average each $10 bet results in an expected profit of $(1/2)(\$10) + (1/2)(-\$11) = \$-.50$. Thus a bettor loses on average $0.50/ $10.50 or $1/21 per dollar bet. Assuming we bet the same amount on each game, to break even we would have to win a fraction p of our bets where $p(10) + (1 - p)(-11) = 0$. The value $p = 11/21 = .524$ solves this equation. Thus to win money on average we must beat the point spread at least 52.4% of the time.

Most bookmakers have power ratings on NFL and NBA teams.[1] These ratings can be used to set point spreads for which the favorite has approximately a 50% chance of covering the spread. For example, if the Colts have a power rating of $+10$ and the Browns have a power rating of -4, we would expect on a neutral field the Colts to win by $10 - (-4) = 14$ points. Of course, teams play better at home. Home edges for various sports (based on the last ten years) are 3 points for the NFL, NBA, and college football and 4 points for NCAA men's basketball. We will see later in the chapter how to estimate the home edge for a given set of games. Using the NFL home edge of 3 points the bookies would favor the Colts by $14 + 3 = 17$ points at home and favor the Colts by $14 - 3 = 11$ points at Cleveland. Predictions created from power ratings usually create (in the absence of injuries) a "fair line" in the sense that the favorite and underdog have an equal chance of covering the prediction.

---

[1] In baseball you bet on a team to win. The probability of a baseball team winning depends heavily on the starting pitchers. We will ignore baseball in our discussions because of this added complexity.

We can now use the simple "point spread setting" system described in the last paragraph to fit power ratings to a set of game results. The file Nfl2006ratings.xls contains the scores of all games (except for the Super Bowl) during the 2006 NFL season. We will show how to determine power ratings for each team and estimate the NFL home edge for the 2006 season. We will constrain our ratings so they average to 0. Thus a team with a power rating of $+5$ is five points better than average while a team with a power rating of -7 is 7 points worse than average. Figure 40.1 contains final season (pre–Super Bowl) power ratings while figure 40.2 excerpts the data for the season's first twelve games. We determine NFL power ratings and league home edge according to the following six steps.

	C	D	E
	Team	Rating	Rank
3	1 Arizona Cardinals	−6.7842786	29
4	2 Atlanta Falcons	−3.408071	21
5	3 Baltimore Ravens	8.3506952	4
6	4 Buffalo Bills	2.4201424	12
7	5 Carolina Panthers	−3.059493	20
8	6 Chicago Bears	8.6780273	3
9	7 Cincinnati Bengals	3.7895627	7
10	8 Cleveland Browns	−6.2649554	28
11	9 Dallas Cowboys	3.283987	8
12	10 Denver Broncos	1.3025094	14
13	11 Detroit Lions	−6.1001288	27
14	12 Green Bay Packers	−4.2083929	25
15	13 Houston Texans	−4.2734808	26
16	14 Indianapolis Colts	7.4756531	6
17	15 Jacksonville Jaguars	7.6999228	5
18	16 Kansas City Chiefs	0.4792909	16
19	17 Miami Dolphins	0.9335118	15
20	18 Minnesota Vikings	−3.8241017	23
21	19 New England Patriots	10.925026	1
22	20 New Orleans Saints	2.777193	11
23	21 New York Giants	0.1635728	17
24	22 New York Jets	1.6926229	13
25	23 Oakland Raiders	−9.7520115	32
26	24 Philadelphia Eagles	3.1500213	9
27	25 Pittsburgh Steelers	3.0247143	10
28	26 St. Louis Rams	−3.7860389	22
29	27 San Diego Chargers	9.9129332	2
30	28 San Francisco 49ers	−8.6147765	31
31	29 Seattle Seahawks	−2.555432	19
32	30 Tampa Bay Buccaneers	−8.2410296	30
33	31 Tennessee Titans	−1.0912204	18
34	32 Washington Redskins	−4.0959744	24

Figure 40.1. NFL power ratings for 2006.

	A	B	C	D	E	F	G	H	I
39									**SSE** **38306.43**
40	Game #	Home	Away	Home score	Away score	Margin	Prediction	Error	Squared error
41	1	25	17	28	17	11	3.0440188	7.96	63.29764
42	2	19	4	19	17	2	9.4576993	−7.5	55.61728
43	3	21	14	21	26	−5	−6.359264	1.36	1.847599
44	4	5	2	6	20	−14	1.3013943	−15	234.1327
45	5	15	9	24	17	7	5.3687521	1.63	2.66097
46	6	8	20	14	19	−5	−8.089332	3.09	9.543973
47	7	30	3	0	27	−27	−15.63891	−11	129.0744
48	8	11	29	6	9	−3	−2.59188	−0.4	0.166562
49	9	31	22	16	23	−7	−1.831027	−5.2	26.71828
50	10	12	6	0	16	−26	−11.9336	−14	197.8635
51	11	26	10	18	10	8	−4.135732	12.1	147.276
52	12	16	7	10	23	−13	−2.357456	−11	113.2638

Figure 40.2. Excerpt of NFL 2006 games' database.

Step 1. Enter trial ratings (in range D3:D34) and a trial home edge (in cell G1) (see figure 40.1). Note each team has a code number listed in cells B3:B32. For example, the Bears are team 6.

Step 2. In figure 40.2, column A gives the game number, column B the code number for the home team, column C the code number for the away team, column D the points scored by the home team, and column E the points scored by the away team. For example, in the first game Pittsburgh was home against Miami and won 28–17.

Step 3. In column F of figure 40.2, we determine the number of points by which the home team won. In game 1 Pittsburgh won by $28 - 17 = 11$ points. Note a negative number in column F means the home team lost. For example, in game 3 the Giants lost at home to the Colts by 5 points.

Step 4. We now generate a prediction (based on our power ratings and home edge) for the home margin of each game. The prediction (implemented using the Excel VLOOKUP formula) is simply (home edge) + (home team rating) − (away team rating). For game 1 our prediction (based on ratings given in figure 40.1) would be for Pittsburgh to win by $0.95 + (3.02) - (.93) = 3.04$ points.

Step 5: In column H (figure 40.2) we compute our forecast error (actual home margin − predicted home margin) for each game. For example in game 1 the home team won by 11 and our prediction was that they would win by 3.04. Therefore our forecast error was $11 - 3.04 = 7.96$ points. Note that a positive forecast error means the home team did better than predicted while a negative error means that the home team did worse than predicted. Intuitively, it seems like the sum of the forecast errors should be 0. This would imply that on average we over forecast by as much as we under forecast. This is indeed the case.

Step 6. In column I (figure 40.2) we square each error. Then we use the Excel Solver to change the home edge and team ratings to minimize the sum of the squared forecast errors. We will constrain the average team rating (computed in cell D1 of figure 40.1) to equal 0. Note that minimizing the sum of squared errors ensures that positive and negative errors do not cancel each other out. (All regression equations estimated in this book were also computed by minimizing squared errors.)

When the Solver completes its magic we can be sure that the team ratings and home edge shown in figure 40.1 do a better job of fitting the scores than does any other set of ratings and home edge. Note the top five teams are the Patriots (10.92 points better than average), Chargers (9.91 points better than average), Bears (8.68 points better than average), Ravens (8.35 points better than average), and Jaguars (7.70 points better than average). The Super Bowl Champion Colts are sixth best. The reason the Colts do not show up that well is that minimizing the sum of squared errors gives a lot of weight to outliers or atypical games. This is because squaring errors magnifies the effects of huge errors on our objective or Target Cell. For example, Jacksonville beat Indianapolis 44–17 at home during 2006. This rout makes it difficult for the computer to rate the Colts ahead of Jacksonville. If we delete this game and rerun our model we find the Colts are now rated at 8.83 and Jaguars at 6.03. Thus this one game resulted in a three-point swing in our estimate of the relative merits of these teams.

Evaluating Strength of Schedule

We can easily use our ratings to calculate the schedule strength faced by each team. We simply average the ability of all opponents played (using Excel's SUMIF function). Figure 40.3 gives the schedule strength faced by each team. The Titans faced the toughest schedule (3.66 points tougher than average) and the Seahawks the easiest schedule (3.12 points worse than average).

Ranking Teams Based on Mean Absolute Errors

We stated earlier that minimizing squared errors gives lots of weight to games with unexpected outcomes like that of the Colts-Jaguars game. As

	C	I	J
2	Team	Mean strength	Rank
3	1 Arizona Cardinals	−2.09677832	29
4	2 Atlanta Falcons	−1.15809019	25
5	3 Baltimore Ravens	−0.47256958	20
6	4 Buffalo Bills	3.107639211	2
7	5 Carolina Panthers	−0.87202078	21
8	6 Chicago Bears	−2.63199544	30
9	7 Cincinnati Bengals	1.164556765	8
10	8 Cleveland Browns	1.110025495	9
11	9 Dallas Cowboys	−1.03560587	24
12	10 Denver Broncos	0.427537164	15
13	11 Detroit Lions	−0.28764014	19
14	12 Green Bay Packers	−0.14590097	18
15	13 Houston Texans	1.914040834	4
16	14 Indianapolis Colts	1.764714034	5
17	15 Jacksonville Jaguars	1.637426141	6
18	16 Kansas City Chiefs	−0.08001327	17
19	17 Miami Dolphins	2.370991586	3
20	18 Minnesota Vikings	−1.01160498	22
21	19 New England Patriots	1.221306556	7
22	20 New Orleans Saints	−1.92742082	28
23	21 New York Giants	0.542383564	13
24	22 New York Jets	1.05604195	10
25	23 Oakland Raiders	0.498006239	14
26	24 Philadelphia Eagles	−1.01503279	23
27	25 Pittsburgh Steelers	0.649710063	11
28	26 St. Louis Rams	−2.91101974	31
29	27 San Diego Chargers	−1.7157506	27
30	28 San Francisco 49ers	−1.48975993	26
31	29 Seattle Seahawks	−3.12246993	32
32	30 Tampa Bay Buccaneers	0.633967294	12
33	31 Tennessee Titans	3.658795494	1
34	32 Washington Redskins	0.216530976	16

Figure 40.3. NFL 2006 team schedule strengths.

an alternative, we can simply take the absolute value of the error for each game and minimize the sum of absolute errors. Thus if in one game the home team wins by 5 points more than expected and in another game the home team wins by 5 points fewer than expected, these games would contribute $|5| + |-5| = 10$ points to our Target Cell. Note that with an absolute value criterion, positive and negative errors do not cancel out. Minimizing the sum of absolute errors gives less weight to unusual games and more weight to a team's typical performance. Figure 40.4 shows the ranking of NFL teams using the minimization of absolute errors criteria. Note the Bears replace the Chargers at #2. In addition, using absolute errors indicates the Raiders were even worse than previously estimated (−12.75 points). Absolute errors indicate that the Tampa Bay Buccaneers

	C	D	E	F	G
2	Team	Absolute	Least squares	Difference	Absolute rank
3	1 Arizona Cardinals	−4.80768	−6.78427861	1.9765953	26
4	2 Atlanta Falcons	−4.36135	−3.40807102	0.9532836	25
5	3 Baltimore Ravens	7.40581	8.35069519	0.9448829	5
6	4 Buffalo Bills	2.80086	2.42014244	0.3807207	9
7	5 Carolina Panthers	−0.34713	−3.05949297	2.7123645	19
8	6 Chicago Bears	10.6558	8.6780273	1.9777636	2
9	7 Cincinnati Bengals	2.40617	3.7895627	1.3833966	10
10	8 Cleveland Browns	−6.11618	−6.26495543	0.1487792	28
11	9 Dallas Cowboys	4.39876	3.28398695	1.1147778	7
12	10 Denver Broncos	0.29567	1.3025094	1.0068357	17
13	11 Detroit Lions	−8.3572	−6.10012879	2.25707	29
14	12 Green Bay Packers	−4.20271	−4.20839289	0.005679	24
15	13 Houston Texans	−2.60321	−4.27348081	1.6702682	22
16	14 Indianapolis Colts	6.41057	7.4756531	1.0650848	6
17	15 Jacksonville Jaguars	8.15662	7.6999228	0.4566968	4
18	16 Kansas City Chiefs	−0.10366	0.47929089	0.5829509	18
19	17 Miami Dolphins	1.15297	0.93351183	0.2194597	13
20	18 Minnesota Vikings	−1.10172	−3.82410173	2.7223776	21
21	19 New England Patriots	12.9031	10.9250255	1.9780502	1
22	20 New Orleans Saints	1.50793	2.77719299	1.2692639	11
23	21 New York Giants	−0.34744	0.16357281	0.5110081	20
24	22 New York Jets	4.15903	1.69262285	2.4664096	8
25	23 Oakland Raiders	−12.8486	−9.75201148	3.096581	32
26	24 Philadelphia Eagles	0.89643	3.15002133	2.2535913	14
27	25 Pittsburgh Steelers	0.89865	3.02471433	2.1260607	14
28	26 St. Louis Rams	−3.05758	−3.7860389	0.7284634	23
29	27 San Diego Chargers	9.15375	9.91293324	0.7591804	3
30	28 San Francisco 49ers	−8.6091	−8.61477653	0.0056805	30
31	29 Seattle Seahawks	0.68754	−2.55543204	3.2429698	16
32	30 Tampa Bay Buccaneers	−12.6028	−8.2410296	4.3617745	31
33	31 Tennessee Titans	1.42194	−1.09122042	2.5131588	12
34	32 Washington Redskins	−5.84522	−4.09597443	1.7492501	27

Figure 40.4. NFL 2006 team ratings, based on absolute errors.

were nearly as bad as the Raiders. The difference column gives the absolute value of the difference between the least squares and absolute value ratings.

Evaluating Team Offenses and Defenses

Bookmakers also allow you to bet on the total points scored in a game. This is called the over/under number. For example, in Super Bowl XLI the over/under was 49.5 points. This means that if you bet over you win if 50 or more points were scored and if you bet under you win if 49 or fewer points are scored. The final score was Colts over Bears 29–17, so the under bet would have won. We can obtain total points predictions by computing an offensive and defensive rating for each team. (See worksheet offdef.) Our changing cells are the home edge, an average number of points scored by a team in a game, and each team's offensive and defensive rating. We will constrain the average of all team offensive and defensive ratings to equal 0. A positive offensive rating means a team scores more points than average while a negative offensive rating means a team scores fewer points than average. A positive defensive rating means a team gives up more points than average while a negative defensive rating means a team gives up fewer points than average.

We set up predictions for the number of points scored by the home and away teams in each game. Then we choose our changing cells to minimize the sum over all games of the squared errors made in predicting the home and away points scored. We predict the points scored by the home team in each game to be

(league average constant) + .5(home edge) + (home team offensive rating) + (away team defensive rating).

We predict the number of points scored by the away team in each game to be

(league average constant) − .5(home edge) + (away team offensive rating) + (home team defensive rating).

Note we divide up the home edge equally among the home team's points and the away team's points.

After minimizing the sum of the squared prediction errors for home and away points scored in each game, we obtain the results shown in figure 40.5. For example, we find that the Chargers had the best offense, scoring 9.69 more points than average. The Ravens had the best defense, yielding 7.96 points less than average. Oakland had the worst offense, scoring 10.34 fewer points than average, while the 49ers had the worst defense, yielding 5.21 more points than average. For each team, the team's overall rating = team offense rating − team defense rating.

We found the average points to be 20.65. Therefore, for the Super Bowl (played on a neutral field), we would have predicted the Colts to score 20.65 + 7.28 − 3.12 = 24.81 points and the Bears to score 20.65 + 5.56 − .20 = 26.01 points. We would have set the over/under total to 24.81 + 26.01 = 50.82, or 51 points (compared to the Vegas over/under total of 49.5 points).

	C	D	E	F	G	H
2	Team	Rating	Offense	Offense rank	Defense	Defense rank
3	1 Arizona Cardinals	−6.784	−2.536	23	4.24852	29
4	2 Atlanta Falcons	−3.408	−2.973	24	0.43532	18
5	3 Baltimore Ravens	8.3507	0.3933	15	−7.95768	1
6	4 Buffalo Bills	2.4201	−0.146	16	−2.56611	6
7	5 Carolina Panthers	−3.059	−4.326	29	−1.26702	8
8	6 Chicago Bears	8.678	5.5568	4	−3.12101	5
9	7 Cincinnati Bengals	3.7896	3.9425	7	0.15263	16
10	8 Cleveland Browns	−6.265	−4.658	30	1.60629	24
11	9 Dallas Cowboys	3.284	4.6149	5	1.331	22
12	10 Denver Broncos	1.3025	−0.738	18	−2.04129	7
13	11 Detroit Lions	−6.1	−1.715	20	4.38525	30
14	12 Green Bay Packers	−4.208	−2.257	22	1.95122	26
15	13 Houston Texans	−4.273	−3.097	25	1.17686	21
16	14 Indianapolis Colts	7.4757	7.2799	2	−0.1957	13
17	15 Jacksonville Jaguars	7.6999	2.7625	10	−4.93731	3
18	16 Kansas City Chiefs	0.4793	−0.336	17	−0.81507	10
19	17 Miami Dolphins	0.9335	−3.307	26	−4.2399	4
20	18 Minnesota Vikings	−3.824	−3.613	28	0.2115	17
21	19 New England Patriots	10.925	5.5767	3	−5.3481	2
22	20 New Orleans Saints	2.7772	4.4639	6	1.68667	25
23	21 New York Giants	0.1636	1.1553	12	0.99185	20
24	22 New York Jets	1.6926	0.4928	14	−1.19964	9
25	23 Oakland Raiders	−9.752	−10.34	32	−0.58763	11
26	24 Philadelphia Eagles	3.15	3.0384	8	−0.11146	15
27	25 Pittsburgh Steelers	3.0247	2.8827	9	−0.14231	14
28	26 St. Louis Rams	−3.786	0.8919	13	4.67792	31
29	27 San Diego Chargers	9.9129	9.6946	1	−0.21857	12
30	28 San Francisco 49ers	−8.615	−3.403	27	5.21129	32
31	29 Seattle Seahawks	−2.555	−1.1	19	1.4557	23
32	30 Tampa Bay Buccaneers	−8.241	−7.332	31	0.90868	19
33	31 Tennessee Titans	−1.091	1.1769	11	2.26823	28
34	32 Washington Redskins	−4.096	−2.046	21	2.04983	27

Figure 40.5. NFL 2006 offensive and defensive team ratings.

Ranking Teams Based Solely on Wins and Losses

College football rankings have long sparked controversy. Each season the BCS (Bowl Championship Series) picks two teams to play for the college football championship. A key element in the BCS final ranking is an averaging of several computer rankings of college football teams. The BCS has dictated that the scores of games cannot be used to rank teams. Only a team's win-loss record can be used to rank teams (see chapter 49 for further discussion of the BCS). The BCS believes allowing game scores to influence rankings would encourage the top teams to run up the score. Using a technique similar to logistic regression (described in chapter 21 when we tried to predict the chance of making a field goal based on the yard line), we can rate NFL teams based simply on their record of wins and losses. Our changing cells will be a rating for each team and a home edge cell. We assume the probability p of the home team winning the game can be determined from

$$\text{Ln} \frac{p}{1-p} = \text{home rating} - \text{away rating} + \text{home edge.} \qquad (1)$$

Rearranging equation (1) we find

$$p = \frac{e^{\text{home rating} - \text{away rating} + \text{home edge}}}{e^{\text{home rating} - \text{away rating} + \text{home edge}} + 1}. \qquad (2)$$

To estimate the ratings for each team we use the method of maximum likelihood. We simply choose the team ratings and home edge that maximize the probability of the actual sequence of wins and losses we observed. Suppose we want to estimate Shaquille O'Neal's chance of making a free throw and we observe that he makes 40 out of 100 free throws. Let p = probability Shaq makes a free throw. Then $1 - p$ is probability Shaq misses a free throw. The probability of observing 40 made and 60 missed free throws is $\frac{100!}{60! \times 40!} p^{40}(1-p)^{60}$. The constant term is simply the number of ways to place 40 made free throws in a sequence of 100 free throws. To maximize this probability, simple calculus (or the Excel Solver) shows we should set $p = .40$. Thus we estimate Shaq is a 40% foul shooter. This estimate certainly agrees with our intuition. In a similar fashion we choose the team ratings and home edge to maximize the following product: for each game won by the home team take the probability the home team wins. For each game won by the away team take $(1 - \text{probability home team wins})$. Multiply these probabilities for each game and choose team ratings and home edge to maximize the product of these probabilities.[2] Figure 40.6 shows the team ratings and home edge using the Excel Solver to perform this optimization.

Using win-loss rankings, the Colts are now the best team. The Chargers are second, the Patriots third, the Bears fourth, and the Ravens fifth.

---

[2] Actually we have the Solver maximize the sum of the natural logarithms of these probabilities. This is computationally convenient and is equivalent to maximizing the product of the probabilities.

	C	D	E
2	Team	Rating	Rank
3	1 Arizona Cardinals	−1.08719	28
4	2 Atlanta Falcons	−0.6461	25
5	3 Baltimore Ravens	1.31922	5
6	4 Buffalo Bills	0.23969	12
7	5 Carolina Panthers	−0.29729	21
8	6 Chicago Bears	1.51406	4
9	7 Cincinnati Bengals	0.14402	14
10	8 Cleveland Browns	−1.17678	29
11	9 Dallas Cowboys	−0.05011	17
12	10 Denver Broncos	0.43278	8
13	11 Detroit Lions	−1.58706	31
14	12 Green Bay Packers	−0.01896	16
15	13 Houston Texans	−0.38187	23
16	14 Indianapolis Colts	1.78109	1
17	15 Jacksonville Jaguars	0.33287	10
18	16 Kansas City Chiefs	0.19421	13
19	17 Miami Dolphins	−0.19562	20
20	18 Minnesota Vikings	−0.67509	26
21	19 New England Patriots	1.61102	3
22	20 New Orleans Saints	0.30923	11
23	21 New York Giants	−0.05129	18
24	22 New York Jets	0.69034	6
25	23 Oakland Raiders	−2.03161	32
26	24 Philadelphia Eagles	0.37106	9
27	25 Pittsburgh Steelers	−0.06076	19
28	26 St. Louis Rams	−0.31736	22
29	27 San Diego Chargers	1.69567	2
30	28 San Francisco 49ers	−0.46343	24
31	29 Seattle Seahawks	0.0179	15
32	30 Tampa Bay Buccaneers	−1.22403	30
33	31 Tennessee Titans	0.49607	7
34	32 Washington Redskins	−0.88467	27

Figure 40.6. NFL 2006 ratings based on wins and losses. See worksheet logit.

The Lions are the next-to-worst team and the Raiders are the worst team. Using these ratings, how would we predict the chances of the Colts beating the Bears in the Super Bowl? Using equation (2) we would estimate the chances of the Colts beating the Bears (remember there is no home edge in the Super Bowl) as $\dfrac{e^{1.781-1.514}}{e^{1.781-1.514}+1}=.566$. Therefore, we estimate the Colts had around a 57% chance to win the Super Bowl. To use another example, if the Colts hosted the Patriots we would estimate the Colts' chance of winning as $\dfrac{e^{1.781-1.611+.219}}{e^{1.781-1.611+.219}+1}=.596$.

The astute reader has probably anticipated a possible problem with rating teams based solely on their wins and losses. Suppose Harvard is the only undefeated college team and the usual powers like USC, Florida, and Ohio State are 11–1. Almost nobody would claim that Harvard is the best team. They simply went undefeated by facing a relatively easy Ivy League schedule. If we run our win-loss ranking system, a sole undefeated team will have an infinite rating. This is unreasonable. Mease came up with an easy solution to this problem.[3] Introduce a fictitious team (call it Faber College) and assume that each real team had a 1–1 record against Faber. Running our win-loss system usually gives results that would place a team like Harvard behind the traditional powerhouses.

We note that our points based and win-loss based rating systems can easily be modified to give more weight to more recent games. Simply give a weight of 1 to the most recent week's games, a weight of λ to last week's games, a weight λ^2 to games from two weeks ago, and so forth. Here λ must be between 0 and 1. For pro football, $\lambda = .95$ seems to work well, while for college football, $\lambda = .9$ works well. Essentially $\lambda = .9$ means last week's game counts 10% less than this week's game. The value of λ can be optimized to give the most accurate forecasts of future games. The Sagarin ratings are generally considered the best set of team rankings. They incorporate a proprietary weighted least squares algorithm.[4]

Ranking World Cup Soccer Teams

This section discusses how to use the scores of the games in the 2006 World Cup to rate the participating countries.

The teams played schedules of varying difficulty so it is not clear that the best team won. For example, we will soon see that group 7 (Korea, Switzerland, Togo, and France) was more than two goals better than group 3 (Argentina, Serbia, Ivory Coast, and the Netherlands).

Our changing cells in sheet ratings are the ratings for each team. Figure 40.7 shows a subset of the game results. (Our choice of the home team is arbitrary.) Figure 40.8 shows the final ratings.

We begin by entering trial ratings in B4:B35. In cell B36 we average these ratings and constrain the average rating to equal 0. This ensures that a better-than-average team has a positive rating and a worse-than-average team has a

[3] Mease, "A Penalized Maximum Likelihood Approach for the Ranking of College Football Teams Independent of Victory Margins."

[4] Sagarin's ratings for the current season of pro and college football and basketball may be found at http://www.usatoday.com/sports/sagarin.htm. Sagarin ratings for past seasons can be found at http://www.usatoday.com/sports/sagarin-archive.htm.

	F	G	H	I	J	K	L	M
1								72.78642
2	Game	Home	Away	Home score	Away score	Margin	Forecasted margin	Squared error
3	1	Germany	Costa Rica	4	4	2	3.129490701	1.27575
4	2	Poland	Ecuador	0	0	−2	−0.87050872	1.27575
5	3	England	Paraguay	1	1	1	0.870508014	0.01677
6	4	Trinidad	Sweden	0	0	0	−1.12949087	1.27575
7	5	Argentina	Ivory Coast	2	2	1	2.040205281	1.08203
8	6	Serbia	Netherlands	0	0	−1	−2.45979481	2.131
9	7	Mexico	Iran	3	3	2	1.290205901	0.50381
10	8	Angola	Portugal	0	0	−1	−1.20979364	0.04401
11	9	USA	Czech Rep.	0	0	−3	−0.74999972	5.0625
12	10	Italy	Ghana	2	2	2	1.402921811	0.3565
13	11	Australia	Japan	3	3	2	1.284793744	0.51152
14	12	Brazil	Croatia	1	1	1	1.71520468	0.51152
15	13	S. Korea	Togo	2	2	1	0.999999619	1.5E−13
16	14	France	Switzerland	0	0	0	0.358982179	0.12887
17	15	Germany	Poland	1	1	1	2.129490291	1.27575
18	16	Spain	Ukraine	4	4	4	1.025649008	8.84676
19	17	Tunisia	Saudi Arabia	2	2	0	0.500000202	0.25
20	18	Ecuador	Costa Rica	3	3	3	1.870509133	1.27575
21	19	England	Trinidad	2	2	2	1.870508282	0.01677
22	20	Sweden	Paraguay	1	1	1	0.129490605	0.75779
23	21	Argentina	Serbia	6	6	6	3.790205797	4.88319
24	22	Netherlands	Ivory Coast	2	2	1	0.709794289	0.08422

Figure 40.7. World Cup 2006 game scores. See file Worldcup06.xls.

negative rating. Our prediction for the margin by which the team on the left (called home team) wins game is home rating − away rating. This implies that a team with a +1 rating is predicted to beat a team with a −0.5 goal rating by 1.5 goals. Then we try to minimize squared errors in the following way.

Step 1. Compute home margins by copying from K3 to K4:K66 the formula I3 − J3.

Step 2. Compute forecast for each game's home margin by copying from L3 to L4:L66 the formula

VLOOKUP(G3,look,2,FALSE) − VLOOKUP (H3,look,2,FALSE).

Step 3. Compute squared forecast error for each game by copying from M3 to M4:M66 the formula (K3 − L3) ^ 2.

Step 4. Compute sum of squared errors (SSE) in cell M1 with the formula SUM(M3:M66).

Step 5. Set up the Excel Solver to minimize squared errors by changing ratings and constraining average ratings to equal 0.

	A	B	C	D
3	**Team**	**Rating**	**Rank**	**Group**
4	Angola	−0.925	24	4
5	Argentina	0.776	11	3
6	Australia	1.1289	5	6
7	Brazil	2.5593	1	6
8	Costa Rica	−2.032	31	1
9	Croatia	0.8441	10	6
10	Czech Rep.	0.6917	12	5
11	Ecuador	−0.161	19	1
12	England	0.3207	13	2
13	France	2.1143	3	7
14	Germany	1.0977	6	1
15	Ghana	0.8652	9	5
16	Iran	−1.675	29	4
17	Italy	2.2681	2	5
18	Ivory Coast	−1.264	27	3
19	Japan	−0.156	18	6
20	Mexico	−0.385	20	4
21	Netherlands	−0.554	23	3
22	Paraguay	−0.55	22	2
23	Poland	−1.032	25	1
24	Portugal	0.2848	14	4
25	S. Korea	0.9348	8	7
26	Saudi Arabia	−1.7	30	8
27	Serbia	−3.014	32	3
28	Spain	1.063	7	8
29	Sweden	−0.42	21	2
30	Switzerland	1.7553	4	7
31	Togo	−0.065	17	7
32	Trinidad	−1.55	28	2
33	Tunisia	−1.2	26	8
34	Ukraine	0.0374	15	8
35	USA	−0.058	16	5
36	Mean rating	2E−15		

Figure 40.8. World Cup 2006 soccer ratings. See file Worldcup06.xls.

The Solver window is shown in figure 40.9.

Although Brazil did not win, they played the best overall, ranking 2.56 goals better than average (see figure 40.8). Italy was second, and France third. Figure 40.10 shows the average ability of each group. We computed each group average by copying from C39 to C40:C47 the formula SUMIF(Group,B40,Rating)/4. This formula averages the ratings of the four teams in each group. As pointed out previously, group 7 was the toughest and group 3 was the worst. Group 7 was around 2.2 goals tougher than group 3. FIFA (which runs World Cup soccer) supposedly attempts to equalize the strengths of the groups, but they certainly failed to do this in 2006.

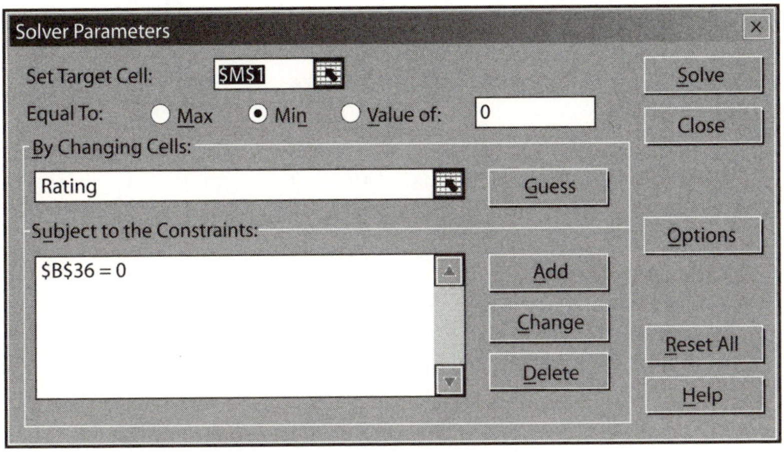

Figure 40.9. Excel Solver window.

	B	C
39	Group	Mean
40	1	−0.532
41	2	−0.55
42	3	−1.014
43	4	−0.675
44	5	0.9417
45	6	1.0941
46	7	1.1848
47	8	−0.45

Figure 40.10. Average strength of 2006 World Cup groups.

Ranking World Cup Offenses and Defense

Figure 40.11 shows the offensive and defensive ratings for each World Cup team. A 1.2 offensive rating means a team scores 20% more goals than average against an average defensive team, while a −0.5 defensive rating means a team gives up 50% fewer goals than average against an average offensive team. The key is to forecast the number of goals scored by each team and minimize the sum of the squared errors computed based on predicted goals scored by each team.

Figure 40.11 shows that Brazil had the best offense and Switzerland the best defense. The predicted number of goals scored by a team in a game is (mean) × (team offense rating) × (team defense rating). For example, when

	A	B	C	D	E
1		**Mean goals per game**	1.355506308		
3	**Team**	**Offense**	**Defense**	**Offense rank**	**Defense rank**
4	Angola	0.210448	0.431121862	30	7
5	Argentina	0.909618	0.434178804	16	8
6	Australia	2.520287	0.550567468	2	10
7	Brazil	2.980436	0.164229293	1	4
8	Costa Rica	0.950901	1.5556266	14	27
9	Croatia	0.696413	0.357374714	20	6
10	Czech Rep.	1.272259	0.883490696	9	18
11	Ecuador	1.33206	0.834428016	8	17
12	England	0.776985	0.078155163	19	2
13	France	1.498209	0.185784115	5	5
14	Germany	2.051855	0.680572054	3	12
15	Ghana	1.223601	0.691021291	10	13
16	Iran	0.45791	1.845508361	28	29
17	Italy	1.743996	0.144327747	4	3
18	Ivory Coast	0.492537	2.011333649	26	30
19	Japan	1.151072	0.91441415	11	19
20	Mexico	1.142497	1.579717003	12	28
21	Netherlands	0.239864	0.440607763	29	9
22	Paraguay	0.629558	1.098586648	24	21
23	Poland	0.641418	0.606918172	23	11
24	Portugal	0.913775	0.748039231	15	14
25	S.Korea	1.366075	0.803071558	7	16
26	Saudi Arabia	0.460602	1.405145635	27	26
27	Serbia	0.670435	4.691676979	21	32
28	Spain	1.460699	0.924839357	6	20
29	Sweden	0.160427	0.801140689	31	15
30	Switzerland	0.794197	3.37461E−05	18	1
31	Togo	0.873176	1.138506664	17	22
32	Trinidad	1.35E−05	2.024012868	32	31
33	Tunisia	0.61743	1.320709872	25	24
34	Ukraine	1.114128	1.277143349	13	23
35	USA	0.647117	1.377716476	22	25

Figure 40.11. World Cup offense and defense ratings, 2006. See worksheet offdef.

Germany plays Costa Rica we expect Germany to score $1.35 \times (2.05) \times (1.56) = 4.32$ goals. (The mean was a changing cell in Solver as well as in each team's offense and defense rating.) We found that an average offensive team scored 1.35 goals per game. We did not use additive offense and defense ratings as we did in American football, because this might lead us to predict a negative number of goals scored in a game.

We can use our predicted score to predict the outcome of the game. Events that occur rarely, like the number of car accidents a driver has in a year, the number of defects in a product, and the number of goals scored in a soccer game are often governed by the Poisson random variable (see

	C	D	E	F
2	Mean	1.5	0.6	
3	Goals scored	Probability Team 1 scores this many goals	Probability Team 2 scores this many goals	Probability Team 1 wins with this many goals
4	0	0.22313016	0.54881164	
5	1	0.33469524	0.32928698	0.183684642
6	2	0.25102143	0.09878609	0.220421571
7	3	0.12551072	0.01975722	0.122609499
8	4	0.04706652	0.00296358	0.046908466
9	5	0.01411996	0.00035563	0.014114385
10	6	0.00352999	3.5563	0.003529852
11	7	0.00075643	3.0483	0.000756424
12	8	0.00014183	2.2862	0.00014183
13	9	2.3638	1.5241	2.36383
14	10	3.5457	9.1448	3.54575
15	Game outcome	Probabillity		
16	Tie	0.26008902		
17	Team 1 wins	0.59219385		
18	Team 2 wins	0.14771713		

Figure 40.12. Example of game outcome probability computation from expected game score.

chapter 16). For a Poisson random variable with mean λ, the probability that the random variable equals x is given by $\lambda^x e^{-\lambda}/x!$.

For example, if we predict that a team will score 1.4 goals, there is a probability $1.4e^{-1.4}/2! = .242$ of scoring 0 goals. The Excel formula POISSON(x,mean,False) gives the probability that a Poisson random variable with mean λ takes on the value x. Suppose we are given the predicted number of goals scored by each team. In the worksheet probability we use the following formulas to predict the probability of each team winning a game and of a tie. We assume no team will score more than 10 goals in a game. Then

$$\text{probability of a tie} = \sum_{i=0}^{i=10} (\text{probability team 1 scores i goals})$$
$$\times (\text{probability team 2 scores i goals}).$$

$$\text{probability team 1 wins} = \sum_{i=1}^{i=10} (\text{probability team 1 scores i goals})$$
$$\times (\text{probability team 2 scores} < i \text{ goals}).$$

probability team 2 wins = 1 − probability of tie
− probability team 1 wins.

Figure 40.12 illustrates how to use these formulas to predict the probabilities of various game outcomes. For example, if we predict team 1 to score 1.5 goals and team 2 to score 0.6 goals, there is a 26% chance of a tie, a 59% chance that team 1 wins, and a 15% chance that team 2 wins.

41

WHICH LEAGUE HAS GREATER PARITY, THE NFL OR THE NBA?

Sports fans love the NFL because it seems like there is always a surprise team that wins the Super Bowl or challenges for the championship. For example, who expected Tampa Bay to win the Super Bowl in 2002? NBA fans complain the same teams (such as Detroit and San Antonio) are always on top. It is easy to show that the NFL does indeed exhibit more parity and unexpected team performances than does the NBA.

If a league has a great deal of parity you would expect it to be difficult to predict a team's performance based on their previous year's performance. That is, teams that do poorly one season should have a good shot at being above average the following season, and vice versa. If a league exhibits little parity you would expect that it would be relatively easy to predict a team's performance for one season using the previous year's performance. What metric should we use to measure team performance? The simplest metric would be regular season wins, but we have seen that NFL teams play schedules that differ by 7 or more points in difficulty. A team with a tough schedule and 9-7 record might be much better than a team that went 11-5 with an easy schedule. We will use the final Sagarin rating as our metric for team performance during a season.[1] These ratings include all regular and post-season games and are quite similar to the least squares ratings described in chapter 40. The file Parity.xls contains the ratings for each NFL team and NBA team for five seasons (for the NBA, 2002–3 through 2006–7 seasons and for the NFL, the 2002–6 seasons). We try to predict each team's Sagarin rating during a year based on their previous year's Sagarin rating. For example, for our NBA data we would have the following four data points for the Spurs: (Spurs 2002–3 rating,

---

[1] See http://www.usatoday.com/sports/sagarin.htm.

Spurs 2003–4 rating), (Spurs 2003–4 rating, Spurs 2004–5 rating), (Spurs 2004–5 rating, Spurs 2005–6 rating), and (Spurs 2005–6 rating, Spurs 2006–7 rating). We simply find the best-fitting line to predict the following season's rating based on the previous season's rating. Sagarin ratings average to 20 in the NFL and 90 in the NBA.

Figures 41.1 and 41.2 give the results obtained using the Excel Trend Curve feature. We find that the following year's NFL team performance has a .35 correlation with the team's performance the previous year, while the following year's NBA team performance has a .56 correlation with the team's performance the previous year. This shows that it is much easier to predict an NBA team's performance than an NFL team's performance us-

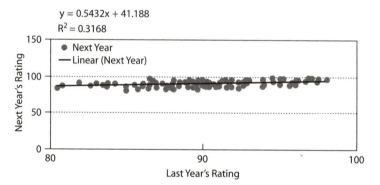

Figure 41.1. Predicting NBA team performance. The correlation between the previous year's Sagarin rating and the following year's rating is obtained by taking the square root of the R Squared values.

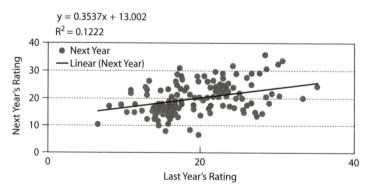

Figure 41.2. Predicting NFL team performance.

ing the previous year's data. This means that good NBA teams are much more likely to stay good than are good NFL teams. Conversely, bad NFL teams have a much better shot at being good the following year than do bad NBA teams. This insight is consistent with the prevalent view that the NFL has more parity than the NBA. Looking at the least squares lines shown in figures 41.1 and 41.2, we would predict that an NBA team that is 10 points better than average during a season (100 rating) would be predicted to have a rating of $41.19 + .5432(100) = 95.51$ the following year (5.51 points better than average), while an NFL team that performed 10 points better than average (30 rating) during a season would be predicted to have a rating of $13 + .3537(30) = 23.61$ (3.61 points better than average). Similarly, an NBA team that is 10 points worse than average would be predicted to play 5.51 points worse than average the following year and an NFL team that played 10 points worse than average would be expected to play only 3.61 points worse than average the following year. Thus for each league we predict a team to play closer to average the following year than they did the previous year. This is an example of regression toward the mean. NFL teams regress more to the mean than do NBA teams. The fact that a good NFL team on average ends up closer to average the following season than does a good NBA team must mean that more good NFL teams play poorly the following season (and more bad NFL teams play well).

Possible Explanations for NFL Parity

Why does the NFL exhibit more parity than the NBA? The NFL has a "hard" salary cap ($109 million per team in 2007), which teams cannot exceed. The NBA penalizes teams that exceed the salary cap ($55.6 million in 2007), but teams are allowed to exceed the salary cap. If an NFL team performs well then their star players may seek higher salaries elsewhere because the hard cap prevents a great team from rewarding all their good players.

In addition, in the NFL draft the worst team is guaranteed the first draft pick. In the NBA a lottery is held and the worst team has only a 25% chance of getting the first pick. In short, the NFL draft has teams draft in inverse order of team performance while the NBA only approximates teams drafting in inverse order of performance. Therefore, it seems reasonable to conclude that the NFL draft would create more parity than would the NBA draft.

The NBA and the NFL also differ in terms of the way player contracts are handled. In the NFL no contract is guaranteed. A player can have a six-year, $100 million contract and if the team cuts the player then the team does not have to pay the player. In the NBA, however, most player contracts are structured so that if a team cuts a player the team still owes the player his money. This dichotomy makes it much easier to change an NFL roster than change an NBA roster. Since it is harder to change an NBA team roster, we would expect that NBA teams' performance during the following season would be more similar to their performance the previous season than would be the case with the performance of an NFL team.

These three differences between the NBA and the NFL appear to be the major contributors to the greater degree of parity exhibited by the NFL.

42

THE RATINGS PERCENTAGE INDEX (RPI)

During the college basketball season hoop fans anxiously anticipate the selection and seeding of teams for the NCAA tournament. The NCAA selection committee wants an accurate view of the teams' relative abilities, but, like the BCS selection committee, the NCAA tournament selection committee wants to use only a team's win-loss record and not the score of their games to rank teams. The NCAA believes that including game scores in the ranking and seeding process would cause the top teams to try and run up the score on lesser opponents. In chapter 40 we explained how to use a logistic regression-based ranking system to rank NFL teams. An identical system would do an excellent job of ranking college basketball teams. Defying logic, however, the NCAA uses the complex and flawed Rating Percentage Index (RPI) to rank college basketball teams.

Let's suppose we want to compute the RPI of Indiana University (IU). IU's RPI ranking is computed as a weighted average of three quantities:

- IU's own winning percentage (referred to as TWP).
- Not counting the games involving IU, the winning percentage of each of IU's opponents. These winning percentages are averaged to compute OPP = opponent's average winning percentage.
- OPPOPP = the average winning probability of IU's opponents' opponents (not including games in which they play any of IU's opponents but including games played against IU).

The home team in college basketball wins around 70% of the games between equally matched teams. This gave teams like Duke, which played many more home games than road games, an unfair advantage in the RPI rankings. Beginning with the 2005 season, a home win or road loss was counted as 0.6 wins or 0.6 losses. An away win or home loss was counted at 1.4 wins or losses. This correction negates the advantage that a team used to receive by dint of playing extra home games.

IU's RPI would now be computed as $.25(TWP) + .50(OPP) + .25(OPPOPP)$. The last two terms in this formula reward a team for playing a schedule including opponents with good winning percentages. Unfortunately, a team can still have a good winning percentage and not be a great team if they play an easy schedule. The RPI formula is a well-intentioned attempt to reward teams for playing a difficult schedule. When discussing Joseph Conrad's *Heart of Darkness*, my twelfth grade English teacher told us all of Conrad's characters had a "fatal flaw." We will soon see that a team can win a game and see their RPI drop, and they can lose a game and see their RPI increase. This counterintuitive property of the RPI is its fatal flaw and should cause the NCAA to throw the baby out with the bath water and replace the RPI with a logistic regression-based rating system. Before detailing the RPI's flaw let's work through a simple example of how the RPI is calculated. For simplicity we will assume that all games are played on a neutral court, so the 1.4 and .6 corrections previously mentioned are not relevant. Here are the relevant results for IU, Purdue, and Iowa:

- IU is 2–0 and has beaten Purdue and Iowa.
- Purdue is 1–1 and lost to IU but beat a team with a 2–1 record.
- Iowa is 1–2. Iowa lost to IU, beat a team with an 0–3 record, and lost to a team with a 3–0 record.

IU's TWP = 1, because IU is 2–0. To compute OPP, note that Purdue is 1–0 without the loss to IU. Iowa is 1–1 without the loss to IU. Therefore $OPP = .5(1 + .5) = .75$.

To compute OPPOPP we need Purdue's and Iowa's opponents' winning percentage. After removing Purdue's results, Purdue's opponents were 1–0 (IU) and 2–0 (other unnamed opponent). Therefore, the average winning percentage of Purdue's opponents is $.5(1 + 1) = 1$. Iowa's three opponents have (excluding their game against Iowa) winning percentages of 1–0 (IU), 0–2, and 2–0. The average win-loss percentage of Iowa's opponents is then $1/3(1 + 0 + 1) = 2/3$. Therefore, IU's OPPOPP $= .5(1 + (2/3)) = .833$. IU's RPI is computed as $.25(1) + .5(.75) + .25(.833) = .8333$.

Jeff Sagarin detected RPI's fatal flaw and has demonstrated that a team can win a game yet see their RPI drop. Suppose our team (IU) has played n games and has won a fraction p of them. In addition, currently our opponents have an average winning percentage OLDOPP. We are playing a new opponent (Navy), whose winning percentage in games not involving IU is given by NAVYWP. We will now compute the change in IU's RPI if they win the game against Navy. We ignore the change in OPPOPP caused by this new game, because this usually has little effect on a team's RPI. After beating Navy,

$$\text{TWP} = \frac{np+1}{n+1} = p + \frac{1-p}{n+1}. \tag{1}$$

Also after beating Navy, OPP will equal

$$\frac{n\text{OLDOPP} + \text{NAVYWP}}{n+1} = \text{OLDOPP}$$
$$+ \frac{\text{NAVYWP} - \text{OLDOPP}}{n+1}. \tag{2}$$

Ignoring the change in OPPOPP the change in IU's RPI is .25(change in TWP) + .50(change in OPP). From (1) and (2) we find that the change in IU's RPI will be less than 0 if and only if

$$\frac{1-p}{4(n+1)} + \frac{\text{NAVYWP} - \text{OLDOPP}}{2(n+1)} < 0. \tag{3}$$

A little algebra shows that (3) is true if and only if NAVYWP $< .5p +$ OLDOPP $- .5$. Suppose $p = .80$ and OLDOPP $= .55$. If NAVYWP $<$ $.5(.8) + .55 - .5 = .45$, then beating Navy will lower IU's RPI.

Similar manipulations show that if IU loses to Navy, then IU's RPI will increase if and only if NAVYWP $> .5p +$ OLDOPP $- .5$. For example, if NAVYWP $= .70$ and IU has won half its games and IU's prior opponents won 40% of their games, then IU's RPI increases after losing to Navy.[1]

---

[1] I hope the NCAA will eliminate this eyesore from the beautiful landscape of college basketball. Current RPI ratings are available at http://kenpom.com/rpi.php.

43

FROM POINT RATINGS TO PROBABILITIES

In chapter 40 we learned how to calculate "power ratings," which allow us to estimate how many points one team is better than another.[1] In this chapter we will show how to use power ratings to determine the probability that a team wins a game, covers a point spread bet, or covers a teaser bet. For NBA basketball, we will see how to use power ratings to determine the probability of each team winning a playoff series. At the end of the chapter we'll look at how power ratings can be used to compute the probability of each team winning the NCAA basketball tournament.

Using power ratings we would predict that the average amount by which a home team will win a game is given by home edge + home team rating − away team rating. The home edge is three points for the NFL, college football, and the NBA. For college basketball the home edge is four points. Stern showed that the probability that the final margin of victory for a home NLF team can be well approximated by a normal random variable margin with mean = home edge + home team rating − away team rating and a standard deviation of 13.86.[2] For NBA basketball, NCAA basketball and college football, respectively, Jeff Sagarin has found that the historical standard deviation of game results about a prediction from a rating system is given by 12, 10, and 16 points, respectively.

Calculating NFL Win and Gambling Probabilities

Let's now focus on NFL football. A normal random variable can assume fractional values but the final margin of victory in a game must be an inte-

---

[1] Even if you do not run the proposed rating systems, you can always look up power ratings at Jeff Sagarin's site on usatoday.com.

[2] Hal Stern, "On the Probability of Winning a Football Game," *American Statistician* 45, no. 3 (August 1991): 179–83

ger. Therefore we estimate the probability that the home team wins by between a and b points (including a and b, where a < b) is probability(margin is between a − .5 and b + .5). The Excel function

$$\text{NORMDIST}(x,\text{mean},\text{sigma},\text{True})$$

gives us the probability that a normal random variable with the given mean and sigma is less than or equal to x. With the help of this great function we can determine the probability that a team covers the spread, wins a game, or beats a teaser bet.

Our power ratings indicate the Colts should win by 7 points. They are only a 3-point favorite. What is the probability that the Colts will cover the point spread?

If the margin is greater than or equal to 3.5, the Colts will cover the spread? If the margin is between 3.5 and 4.5, the game will be a push. The probability the Colts will cover the spread can be computed as

$$1 - \text{NORMDIST}(3.5,7,13.86,\text{TRUE}) = .6.$$

Probability of a push
$$= \text{NORMDIST}(4.5,7,13.86,\text{TRUE})$$
$$- \text{NORMDIST}(3.5,7,13.86,\text{TRUE}) = .028.$$

Therefore, if we throw out the pushes (because no money changes hands), we would expect a bet on the Colts to have a $.6/(1 - .028) = 61.7\%$ chance of covering the spread and a 38.3% chance of losing against the spread.

The Colts are a 7-point favorite in Super Bowl XLI. What is the probability that they will win the game?

Here we assume the point spread equals the mean outcome of the game. The Colts can win with a final margin of 1 point or more or win with, say, a .5 probability if regulation time ends in a tie. The probability the Colts win by 1 or more =

$$1 - \text{NORMDIST}(0.5,7,13.86,\text{TRUE}) = 1 - .3196 = .6804.$$

The probability regulation ends in tie =

$$\text{NORMDIST}(.5,7,13.86,\text{TRUE}) - \text{NORMDIST}$$
$$(-.5,7,13.86,\text{TRUE}) = .0253.$$

Therefore, we estimate the Colts' chance of winning Super Bowl XLI to be .6804 + .5 ×(.0253) = .693. Recall that the money line on this game was Colts −250 and Bears +200. With this line you should have bet on the Colts to win if you thought the Colts had at least a 71.4% chance of winning, and you should have bet on the Bears to win if you thought the Colts had a chance of winning that was less than or equal to 66.7%. The average of these two probabilities is 69.1%, which is almost exactly equal to our estimated chance of the Colts winning (69.3%).

The Colts are an 8-point favorite and we have bet a 7-point teaser on the Colts. What fraction of the time can we predict the Colts will cover our new spread (Colts by 8 − 7 = 1 point)?

If the Colts win by 2 points or more we win the teaser. If the Colts win by 1 point the teaser is a push. If regulation ends in a tie and the game goes into overtime we will win the teaser if and only if the Colts win in overtime. We estimate the Colts' chances of winning in overtime to be 0.5. The probability the Colts win by 2 points or more =

$$1 - \text{NORMDIST}(1.5,8,13.86,\text{TRUE}) = .6804.$$

The probability of a push =

$$\text{NORMDIST}(1.5,8,13.86,\text{TRUE}) - \text{NORMDIST} \\ (.5,8,13.86,\text{TRUE}) = .02533.$$

The probability the game goes into overtime =

$$\text{NORMDIST}(0.5,8,13.86,\text{TRUE}) - \text{NORMDIST} \\ (-0.5,8,13.86,\text{TRUE}) = .0243.$$

Therefore, we estimate the fraction of non-push games in which the Colts will cover the teaser to be

$$\frac{.6804+.5\times(.0243)}{1-.02533} = .711.$$

As we pointed out in chapter 31, Wong observed that 70.6% of all 7-point teasers covered the teaser spread. This compares quite well with our estimate of 71.1%. For this particular teaser, a regulation tie gives us a chance to cover. If we were betting on a 15-point favorite with a 7-point teaser, a tie would be a sure loss.

What were the chances that the Spurs would beat the Cavs in the 2007 NBA finals?

We now turn our attention to trying to translate power ratings into an estimate of an NBA team's chance of winning a playoff series. The first team to win four games in an NBA playoff series is the winner. We will use Excel to "play out" or simulate a seven-game series thousands of times and track the probability of each team winning the series. The home margin of victory in an NBA game will follow a normal distribution with mean = home edge + home rating − away rating and sigma = 12. (The sigma of 12 points is the historical standard deviation of actual game scores about a prediction from a ranking system.)

The Excel formula NORMINV(RAND(),mean, sigma) can generate a sample value from a normal random variable with a given mean and sigma. RAND() creates a random number equally likely to assume any value between 0 and 1. Suppose RAND() = p. Then our formula generates the 100p percentile from the given normal random variable. Thus RAND() = .5 yields the 50th percentile, RAND() = .9 yields the 90th percentile, and so forth. To illustrate how to determine the chance of a team winning the playoff series let's look at the 2007 NBA Finals between the San Antonio Spurs and the Cleveland Cavaliers. Running our power rankings as described in chapter 40 we found the following ratings going into the finals: Spurs = 8.28, Cavs = 3.58. We also found that in 2007 the NBA had a home edge of 3.21 points. Therefore, when the Spurs are at home they would be favored to win by $8.28 - 3.58 + 3.21 = 7.91$ points and when the Spurs are on the road they would be favored to win by $8.28 - 3.58 - 3.21 = 1.49$ points. Therefore, we can simulate the final margin (from the Spurs' viewpoint) of the Spurs' home games (games 1, 2, 6, and 7) with the Excel function

$$=NORMINV(RAND(),7.91,12).$$

When Cleveland is at home (games 3, 4, and 5), we simulate the final Spurs' margin with the formula

$$=NORMINV(RAND(),1.49,12).$$

Although the series may not go seven games, we have Excel simulate the result of seven games and declare the Spurs the winner if they win at least

	D	E	F	G	H	I
9	Spurs' Rating	8.28				
10	Cavs' Rating	3.58				
11	Home Edge	3.21				
13			Game	Spurs favored by (pts.)	RAND	Result
14			1	7.91	0.01785	−17.2939
15			2	7.91	0.689457	13.84174
16			3	1.49	0.005901	−28.7261
17			4	1.49	0.014766	−24.6257
18			5	1.49	0.28071	−5.4788
19			6	7.91	0.626176	11.77091
20			7	7.91	0.161885	−3.93088

Figure 43.1. A sample iteration of Spurs-Cavs finals.

four games. We used EXCEL'S Data Table feature (see file Spurscavs82.xls and the chapter appendix) to play out the series two thousand times. In the sample iteration shown in figure 43.1, the Spurs would have lost in 5 games, winning only game 2.

We found the Spurs won 82% of the time. Recall that the money line on this series was Spurs −450, Cavaliers +325. This would imply that a bettor should bet on the Spurs if they believed the Spurs had a chance of winning the series of at least 82% and should bet on the Cavaliers if they believed that the Spurs' chance of winning the series was less than 76%. Our analysis would have leaned toward betting on the Spurs. (The Spurs swept the Cavs.)

Estimating NCAA Tournament Probabilities

We can use a similar methodology to have Excel "play out" the NCAA tournament several thousand times. We recommend using the Sagarin power ratings from *USA Today* and a template to simulate future NCAA tournaments (see file NCAA2007.xls). In row 2 you type in the names of the teams and in row 4 you type in each team's Sagarin rating (printed in *USA Today* the Monday before the tournament starts). The teams listed in the first 16 columns can be from any region. The next 16 teams listed must be from the region that plays the first listed region in the semifinal games. Then the next two regions may be listed in any order. Within each region the teams should be listed in the order in which they appear in any bracket. Most brackets list the teams according to their seeds in the following order: 1, 16, 8, 9, 4, 13, 5, 12, 3, 14, 6, 11, 7, 10, 2, 15. Listing the ratings

	A	B	C	D	E	F	G	H
2	No. Carolina	E. Kentucky	Marquette	Michigan St.	So. Carolina	Arkansas	Texas	New Mexico St.
3	1	2	3	4	5	6	7	8
4	93.48	71.35	84.35	85.75	84.62	83.9	86.67	78.97
5	Eastern Division							
6	1		2		3		4	
7	26.042	1			18.315	4		
9	1		3		5		7	
10	11.273	1			14.377	7		

Figure 43.2. Simulating the NCAA East Region.

in this order ensures that in the later rounds the winners of the earlier games play the correct opponents. A sample iteration of part of the first round of the 2007 East Region is shown in figure 43.2.

North Carolina (team 1) plays Eastern Kentucky (team 2). The final margin of the game (from the standpoint of the team listed directly above the game outcome) was simulated in cell A7 with the formula NORMINV(RAND(), 93.48 − 71.35,10). Note there is no home court advantage in the NCAA tournament. If a team plays in their own or a neighboring city you might want to give them a 2-point home edge. We found North Carolina won by 26.04 points. Therefore, our spreadsheet puts a 1 in A9 to allow North Carolina to advance to their next game. Similarly, we see team 3 (Marquette) beat team 4 (Michigan State). The winner of the game between team 3 and team 4 (team 3) is advanced to cell C9. Now teams 1 and 3 play. The outcome of this game is simulated with the formula

$$\text{NORMINV(RAND(), 93.48} - 84.35,10).$$

We found team 1 (North Carolina) won by 11. Each region is played out in this fashion. Then in row 58 Excel plays out the NCAA semifinals (see figure 43.3). The spreadsheet "knows" to pick off the winner of each region. We find that Ohio State and Wisconsin win the semifinal games. Then in row 64 we play out the final game. Here Ohio State beat Wisconsin by 7.86 points.

We then used Excel's Data Table feature to play out the tournament a thousand times and Excel's COUNTIF function to count the number of times each team won. Table 43.1 shows the results of this simulation (which teams would have had the greatest chance of winning the tournament). Together we estimated these teams had an 81% chance of winning the tournament. (Florida easily won the tournament.)

	B	C	D	E	F	G	H
57	West		Midwest		East		South
58	1		17		47		57
59	−17.12	17			7.47	47	
60		Winner:				Winner:	
61		Ohio St.				Wisconsin	
62	Finals						
63		17		47			
64		7.86	17				
66			Winner:				
67			Ohio St.				

Figure 43.3. Simulating the Final Four.

TABLE 43.1
Probability of Winning (NCAA)

Team	Probability
North Carolina	18%
Ohio State	15%
Florida	11%
Kansas	11%
Wisconsin	7%
UCLA	7%
Georgetown	7%
Memphis	5%

Game-by-Game Pool Bracket

Each year many people fill in a game-by-game NCAA pool entry. To determine which team should be entered in your pool as the winner of each game, simply track the outcome of each game and choose the team that wins that game most often during our simulation. For example, to pick your pool entry for the East Region, you would track the outcome of the cell (D16), which contains the winner of the East Region. The team that appears in cell D16 the most should be your pick in the pool for that game. You would find that North Carolina wins the East Region much more often than anyone else and thus pick North Carolina as the winner of the East Region.

APPENDIX

Using Data Tables to Perform Simulations

The appendix to chapter 1 reviewed the basics of Excel Data Tables. In the file Spurscavs82.xls we replay the Spurs-Cavs series two thousand times. In cell G19 the formula

$$=IF(COUNTIF(I10:I16,">0") >= 4,1,0)$$

returns a 1 if and only if the Spurs win the series and a 0 if the Cavs win the series. We make cell G19 the output cell for our one-way data table. When we set up the Data Table (the table range is F21:G2021) we make the column input cell any blank cell in the spreadsheet. Then Excel places a 1 in the selected column input cell and computes cell G19. This is one iteration of the Spurs-Cavs series. Then Excel places a 2 in the blank column input cell Then all our =RAND() functions recalculate and we get another "play" or iteration of the Spurs-Cavs series. Therefore our Data Table replays the series two thousand times and we find that the Spurs win around 82% of the time.

44

OPTIMAL MONEY MANAGEMENT

The Kelly Growth Criteria

Suppose we believe we have an almost sure bet on Colts −12. We believe the Colts have a 90% chance of covering the spread. This would probably never happen, but let's assume that such a bet really exists. What fraction of our capital should we allocate to this bet? If we bet all our money many times on bets with a 90% chance of winning, eventually we will be wiped out when we first lose a bet. Therefore, no matter how good the odds, we must be fairly conservative in determining the optimal fraction of our capital to bet.

Edward Kelly determined the optimal fraction of capital to bet on any one gamble.[1] Kelly assumes our goal is to maximize the expected long-run percentage growth of our portfolio measured on a per gamble basis. We will soon see, for example, that if we can pick 60% winners against the spread, on each bet we should bet 14.55% of our bankroll and in the long run our capital will grow by an average of 1.8% per bet. Kelly's solution to determining the optimal bet fraction is as follows. Assume we start with $1. Simply choose the fraction to invest that maximizes the expected value of the natural logarithm of your bankroll after the bet. The file Kelly.xls contains an Excel Solver model to solve for the optimal bet fraction given the following parameters:

- WINMULT = the profit we make per $1 bet on a winning bet.
- LOSEMULT = our loss per $1 bet on a losing bet.
- PROBWIN = probability we win bet.
- PROBLOSE = probability we lose bet.

For a typical football point spread bet, WINMULT = 1 and LOSEMULT = 1.1. For a Super Bowl money line bet on Colts −240,

[1] The story of Kelly's work is wonderfully told in Poundstone, *Fortune's Formula*.

WINMULT = 100/240 = .417 and LOSEMULT = 1. For a Super Bowl money line bet of Bears +220, WINMULT = 220/100 = 2.2 and LOSEMULT = 1. Kelly tells us to maximize the expected value of the logarithm of our final asset position. Given a probability p of a winning bet, we should choose f = fraction our fraction of capital to bet to maximize

$$\text{expected LN final wealth} = p\text{LN} (1 + \text{WINMULT} \times f) + (1 - p)$$
$$\times \text{LN} (1 - \text{LOSEMULT} \times f). \qquad (1)$$

We find the optimal value of f by setting the derivative of (expected final wealth) to 0. This derivative is

$$\frac{p \times \text{WINMULT}}{(1 + \text{WINMULT} \times f)} - \frac{(1-p) \times \text{LOSEMULT}}{(1 - \text{LOSEMULT} \times f)}.$$

This derivative will equal 0 if

$$f = \frac{p\text{WINMULT} - (1-p)\text{LOSEMULT}}{\text{WINMULT} \times \text{LOSEMULT}}.$$

The numerator of the equation for f is our expected profit on a gamble per dollar bet (often called the "edge"). Our equation shows that the optimal bet fraction is a linear function of the probability of winning a bet, a really elegant result.

Kelly also showed that in the long run, betting a fraction f of your bankroll each time leads to a long-term growth rate per gamble of $e^{\text{expected LN final wealth}}$, where expected LN final wealth is given by equation (1).

Simplifying our expression for f we may rewrite our equation for f as

$$f = \frac{p}{\text{LOSEMULT}} - \frac{q}{\text{WINMULT}}.$$

Here we let q = 1 - p represent the probability that we lose the bet. Also an increase in the probability of losing or an increase in LOSEMULT will decrease our bet.

As an example, let's compute our optimal bet fraction for an NFL point spread bet with a 60% chance of winning. We find that

$$f = \frac{.6(1) - .4(1.1)}{1(1.1)} = .145.$$

	M	N	P
	Probability win	**Fraction**	**Expected growth per gamble (%)**
14			
15	0.54	0.0309	0.053
16	0.55	0.0500	0.138
17	0.56	0.0691	0.264
18	0.57	0.0882	0.430
19	0.58	0.1073	0.639
20	0.59	0.1264	0.889
21	0.6	0.1455	1.181
22	0.61	0.1645	1.516
23	0.62	0.1836	1.896
24	0.63	0.2027	2.320
25	0.64	0.2218	2.790
26	0.65	0.2409	3.307
27	0.66	0.2600	3.873
28	0.67	0.2791	4.488
29	0.68	0.2982	5.154
30	0.69	0.3171	5.873
31	0.7	0.3364	6.647
32	0.71	0.3555	7.478
33	0.72	0.3745	8.368
34	0.73	0.3936	9.319
35	0.74	0.4127	10.335
36	0.75	0.4318	11.418
37	0.76	0.4509	12.572
38	0.77	0.4700	13.800
39	0.78	0.4891	15.106
40	0.79	0.5082	16.496
41	0.8	0.5273	17.973
42	0.81	0.5464	19.544
43	0.82	0.5655	21.214
44	0.83	0.5845	22.991
45	0.84	0.6036	24.883
46	0.85	0.6227	26.898
47	0.86	0.6418	29.047
48	0.87	0.6609	31.342
49	0.88	0.6800	33.795
50	0.89	0.6991	36.424
51	0.9	0.7182	39.246

Figure 44.1. Kelly growth strategy and average growth rate
of bankroll as function of win probability.

Figure 44.1 summarizes the optimal bet fraction and expected percentage
growth per gamble using the Kelly growth criteria. Figures 44.2 and 44.3
summarize the dependence of the optimal bet fraction and expected long-
term growth rate on our win probability. As stated earlier, the optimal bet
fraction is a linear function of our win probability, but our average capital
growth rate per gamble increases at a faster rate as our win probability increases.

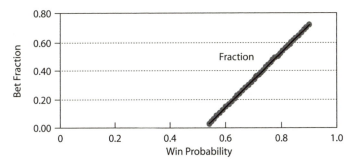

Figure 44.2. Optimal bet fraction as function of win probability.

Figure 44.3. Average wealth growth per period as a function of win probability.

	K	L
	Fraction	**Average growth rate**
2		
3	0.05	0.006668091
4	0.1	0.010628669
5	0.15	0.011796783
6	0.2	0.010058642
7	0.25	0.005266501
8	0.3	−0.002768628
9	0.35	−0.014287401
10	0.4	−0.02959722
11	0.45	−0.049094515
12	0.5	−0.073298724

Figure 44.4. Average long-term growth rate versus fraction bet.

To show the importance of the optimal bet fraction, suppose that we can win 60% of our football point spread bets. Figure 44.4 shows how our long-term average growth rate per bet varies as a function of the fraction bet on each game. If we bet 30% or more of our money on each game, in the long run our capital will decline even though we win 60% of our bets.

RANKING GREAT SPORTS COLLAPSES

With seventeen games left in the 2007 baseball season the New York Mets held a seemingly comfortable seven-game lead in the National League East over the second-place Philadelphia Phillies. The Mets collapsed and the Phillies won the division. This collapse inspired Fox News sportswriter Todd Behrendt to write an article ranking the "all-time great sports collapses."[1] In this chapter we will make some simple assumptions and then use basic probability to try and determine the probability of each collapse occurring. The "greatest collapse" would then be the collapse that had the smallest probability of occurring.

A List of Great Collapses

We now describe the great baseball, basketball, and football collapses listed by Behrendt. (I added the last three collapses on the list.)

- The 2007 Mets (described above).
- On September 20, 1964, the Philadelphia Phillies held a 6.5-game lead over both the St. Louis Cardinals and the Cincinnati Reds. The Phillies went 3–9 down the stretch and the St. Louis Cardinals won the National League title.
- On August 12, 1951, the Brooklyn Dodgers had a 72–36 record and the New York Giants had a 62–51 record. The two teams had seven head-to-head contests remaining. The Giants wiped out the Dodgers' 13.5-game lead and beat the Dodgers in a three-game playoff capped off by Bobby Thompson's amazing home run (see chapter 8).
- The L.A. Lakers trailed the Portland Trailblazers by 15 points with 10:28 left in the fourth quarter of the deciding game 7 of the 2000 NBA Western Conference Finals. The Lakers came back to win.

[1] See http://msn.foxsports.com/mlb/story/7286840.

- The Houston Oilers led the Buffalo Bills 35–3 with 28 minutes left in the 1992 AFC Wild Card Game. The Oilers came back to win.
- The Red Sox were down 3 games to 0 to the New York Yankees during the 2004 American League Championship Series. The Red Sox rallied and won four straight games to take the American League title.
- With two men out and nobody on base, the Mets trailed the Red Sox 5–3 in the bottom of the tenth inning of game 6 of the 1986 World Series. Mookie Wilson's ground ball went through Bill Buckner's legs at first base and the rest is history.
- During game 4 of the 1929 World Series the Philadelphia Athletics trailed the Chicago Cubs 8–0 at the start of the bottom of the seventh inning. The Athletics rallied and won the game.
- With 4:05 left in the Illinois–Arizona 2005 NCAA basketball tournament game, the Arizona Wildcats had a 75–60 lead. Illinois won the game.
- During the 2001 college basketball season Maryland led Duke by 10 points with around a minute left. Duke came back to win the game.

We will now lay out some reasonable assumptions and compute the probability of each "collapse." By far the "winner" as the greatest collapse is the last one, in which Maryland blew a ten-point lead to Duke. (See the file Collapses.xls.)

What Were the Chances the Phillies Would Catch the Mets in 2007?

On September 12, 2007, the Mets had an 83–62 record and the Phillies had a 76–69 record. There were three games remaining between the Mets and Phillies. What were the chances that the Phillies would finish ahead of the Mets and win the National League East? To date the two teams had won the same percentage of their games: $(83 + 76)/(83 + 76 + 62 + 69) = 0.55$. We will therefore assume that in each non-head-to-head game both the Phillies and Mets had a .55 chance of winning, while in head-to-head games both team had a .5 chance of winning. Assuming that the outcomes of successive games are independent events, we used Monte Carlo simulation (see chapter 4) to play out the last 17 games of the season 50,000 times and found the Mets finished in second place 1.2% of the time. In case of the regular season ending in a tie, we assumed each team had a .5 chance to win the playoff.

What Were the Chances the 1964 Phillies
Would Lose the Pennant?

On September 20, 1964, the Phillies had a 90–60 record and held a 6.5-game lead over the Cardinals and Reds, each of whom had an 83–66 record. The Phillies had five games remaining with the Reds and three with the Cardinals. The Cardinals and Reds did not have to play each other. Again we used Monte Carlo simulation to play out the rest of the season. Since all three teams had each won around 57% of their games through September 20, we assumed they all had a .57 chance of winning any non-head-to-head game. We assumed each team also had an equal chance of winning a head-to-head game. In case of a two-way tie we assumed each team had a .5 chance of winning the pennant, and in case of a three-way tie we assumed each team had a 1 in 3 chance of winning the pennant. After using Monte Carlo simulation to play out the remainder of the season 40,000 times, we found that the Phillies had a 1.8% chance of not winning the pennant. This probability may seem higher than expected, but note that the Phillies had a much harder schedule than the Reds or Cardinals.

What Were the Chances the 1951
Giants Would Overtake the 1951 Dodgers?

On August 12, 1951, the Brooklyn Dodgers had a 72–36 record and the New York Giants had a 62–51 record. The two teams had seven head-to-head contests remaining. Since the two teams had each won 60% of their games to date, we assumed that each team had a .6 chance of winning a non-head-to-head-game and a .5 chance of winning a head-to-head game. We found the Giants won the pennant during a fraction 0.0025 (or 1 in 400) of our 40,000 simulations.

What Were the Chances the Blazers Would Blow a
15-Point Fourth-Quarter Lead to the Lakers?

In chapter 43 we stated that the final scoring margin of an NBA game follows a normal random variable with mean = prediction for game outcome and standard deviation = 12 points. If we assume that the changes in margins during different parts of the game are independent and follow the same distribution (the technical term is identically distributed), then the standard deviation of the margin during n minutes of an NBA game is

$$\frac{12}{\sqrt{\dfrac{48}{n}}}^{2}. \tag{1}$$

In general, the standard deviation of the margin of victory during an n minute portion of a game is simply

$$\frac{(\text{game standard deviation of margin})}{\times \sqrt{\text{fraction of game that n minutes is}}}.$$

Therefore, with 10.43 minutes left the standard deviation of the change in game margin is found by substituting $n = 10.43$ in (1). This yields sigma = 5.6.

   Assuming two equally matched teams, we can assume the change in margin has a mean of 0. Then the chance that the Lakers win the game can be approximated with the formula

$$=\text{NORMDIST}(-15.5, 0, 5.6, \text{TRUE}) + 0.5$$
$$\times (-\text{NORMDIST}(-15.5, 0, 5.6, \text{TRUE})$$
$$+ \text{NORMDIST}(-14.5, 0, 5.6, \text{TRUE})).$$

The first term gives the probability the Blazers "win" the last ten minutes by 15.5 points or fewer (and lose the game outright), while the final term computes the probability that the Blazers win the last ten minutes by -15 points and then lose in overtime with probability .5. We find the chance of a successful Lakers comeback was around .004, or one chance in 250.

---

[2] Final margin = margin for minute $1 +$ margin for minute $2 + \ldots$ margin for minute 48. Assuming independence of these random variables, then the variance of the sum is the sum of individual variances so $48 \times$ (variance for one minute) $= 12^2$, or the standard deviation for one minute $= \dfrac{12}{\sqrt{48}}$. Then the variance of margin for n minutes $= n\left(12/\sqrt{48}\right)^2 = \dfrac{144n}{48}$. Finally, the standard deviation of margin for n minutes equals $\dfrac{12}{\sqrt{\dfrac{48}{n}}}$.

What Were the Chances That Houston Would Blow a 35–3 Lead to Buffalo with 28 Minutes Left?

In chapter 43 we stated the outcome of an NFL game follows a normal random variable with mean = game prediction and a standard deviation of 14 points.

Assuming two evenly matched teams, we assume the mean = 0. Then analogous to (1) for an NBA game, we find the standard deviation of the margin during n minutes of an NFL game is

$$\frac{14}{\sqrt{\dfrac{60}{n}}} . \tag{2}$$

For n = 28 we find a standard deviation of 9.56 points. Then assuming the two teams are evenly matched, the chance that Buffalo would come back to win is approximated by =NORMDIST(−32.5,0,9.56,TRUE) + 0.5 × (NORMDIST(−32.5,0,9.56,TRUE) − NORMDIST(−31.5,0,9.56, TRUE)). This yields 3 chances in 10,000, or 1 chance in 3,000, that the Bills would come back and win.

What Were the Chances the Red Sox Would Come Back from Being Down 3 Games to 0 to Win the 2004 American League Championship Series?

If we assume the Red Sox and Yankees to be evenly matched teams and that the outcomes of each game are independent, then the chance of the Red Sox winning four straight games is simply $(.5)^4 = .0625$. This means that given our assumptions, there is a 6% chance that a team will come back and win after they trail a best-of-seven series 3–0. Let's assume the team down 3–0 is down 3–0 because they are inferior. For example, assume the Red Sox have only a 0.25 chance of winning each game. Then their chance of coming back to win the series is $(.25)^4 = .004$, or 4 chances in 1,000. This analysis indicates that the Red Sox comeback was not that unexpected. Behrendt picked this as his #1 collapse. From a probability standpoint, the Yankee collapse was not that unusual, but the breaking of the Red Sox "curse" and the sheer interest in this series may justify the fact that it topped the collapse "charts."

What Were the Chances the Mets Would Win Game 6 of the 1986 World Series before Buckner's Error?

Figure 45.1 shows that by using the Win Expectancy Finder (see chapter 8), with two outs and nobody on in the bottom of the ninth inning the Mets had about a 1.4% chance of winning the game. We used the bottom of ninth inning even though the Mets were batting in the bottom of the tenth inning, because a team batting in the bottom of the ninth has the same chance of winning as a team batting in the bottom of the tenth.

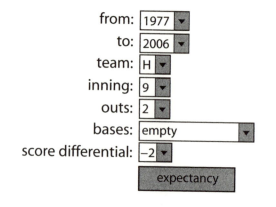

games: 3033

home won: 41

expectancy: 0.014

Figure 45.1. The chance the Mets come back to win game 6 of the 1986 World Series.

What Is the Probability That the Athletics Would Come Back from Being down 8–0 to Beat the Cubs during Game 4 of the 1929 World Series?

Again we consult the Win Expectancy Finder. Figure 45.2 shows that with a 7-run deficit in the bottom of the seventh inning, there are roughly 6 chances in a 1,000 of winning the game. Of course, an 8-run deficit would have a slightly smaller chance of being erased.

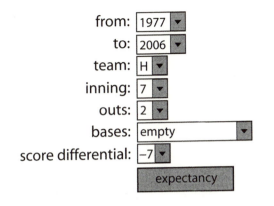

from:	1977 ▼
to:	2006 ▼
team:	H ▼
inning:	7 ▼
outs:	2 ▼
bases:	empty ▼
score differential:	-7 ▼
	expectancy

games: 3053

home won: 17

expectancy: 0.006

Figure 45.2. The chance the Cubs blow an $8-0$ lead in game 4 of the 1929 World Series.

What Is the Probability That Illinois Would Come Back to Beat Arizona?

In chapter 43 we stated that the outcome of an NCAA basketball game follows a normal random variable with mean = game prediction and a standard deviation of 10 points.

Assuming two evenly matched teams, we assume the mean = 0. Then analogous to (1) for an NBA game, we find the standard deviation of the margin during n minutes of an NCAA game is

$$\frac{10}{\sqrt{\dfrac{40}{n}}} . \tag{3}$$

The standard deviation from (3) for a 4:05 time segment is 3.2 points. Therefore, the probability of Illinois coming back to win can be approximated by

$$\text{NORMDIST}(-15.5,0,3.20,\text{TRUE}) + 0.5$$
$$\times (-\text{NORMDIST}(-15.5,0,3.20,\text{TRUE})$$
$$+ \text{NORMDIST}(-14.5,0,3.20,\text{TRUE})).$$

This works out to 4 in 10,000,000.

What Is the Probability That Duke Would Come Back to Beat Maryland?

We will assume the teams are evenly matched. Then in one minute the change in game margin will follow a normal distribution with mean = 0 and a standard deviation of $10/\sqrt{40} = 1.58$ points. The probability Duke will win given they trail by 10 points with a minute left can be computed as

$$1 - \text{NORMDIST}(10.5,0,1.58,\text{TRUE}) + 0.5$$
$$\times (\text{NORMDIST}(10.5,0,1.58,\text{TRUE})$$
$$- \text{NORMDIST}(9.5,0,1.58,\text{TRUE})),$$

with

$$1 - \text{NORMDIST}(10.5,0,1.58,\text{TRUE})$$

is the chance Duke outscores Maryland by more than 10 points in the last minute and

$$0.5 \times (\text{NORMDIST}(10.5,0,1.58,\text{TRUE})$$
$$- \text{NORMDIST}(9.5,0,1.58,\text{TRUE}))$$

is the probability that Duke outscores Maryland by 10 points in the last minute and wins in overtime. This calculates out to less than one chance in a billion.

Our analysis shows that Maryland's collapse had the smallest chance of occurring. As great collapses continue to amaze us, entertain us, and break our hearts, you are now equipped to estimate the likelihood of a collapse.

46

CAN MONEY BUY SUCCESS?

We all know money can't buy love or happiness. In professional sports, can spending more money on players buy a team more success? Let's analyze this question for the NFL, NBA, and MLB.

Does a Larger Payroll Buy Success in the NFL?

In chapter 40 we learned how to calculate offensive and defensive power ratings for NFL teams. For example, an offensive team rating of $+3$ means a team (after adjusting for the strength of opposition) scores 3 points more than average and a defensive rating of -5 means that (after adjusting for the strength of opposition) that a team gives up 5 fewer points than average. For the 2001–4 seasons we tabulated the amount of money each NFL team paid their offensive and defensive personnel. A sample of our data is shown in figure 46.1.

In 2004 NFL players were paid 28.6% more than in 2001, 22.9% more than in 2002, and 7.3% more than in 2003. We would like to predict team offensive performance as a function of total offensive salary and team defensive performance as a function of defensive salary. To ensure all expenditures are measured in 2004 dollars, we must multiply each team's 2001 expenditures by 1.286, each team's 2002 expenditures by 1.229 and each team's 2003 expenditures by 1.073. Using Excel's Trend Curve feature, we can find the straight line that best predicts a team's offensive rating from their offensive players' salaries (in millions of 2004 dollars). The results are shown in figure 46.2.

Offensive salary explains only 6% of offensive team rating and the correlation between offensive salary and team offensive rating is 0.24. Our best straight line equation for predicting offensive team rating is team rating $= -5.4957 + .1556$(offensive team salary in millions). This equation

	B	C	D	E	F	G
8	**Team**	**Year**	**Offensive salary**	**Defensive salary**	**Offensive rating**	**Defensive rating**
9	Arizona	2004	32.6345	34.3735	−5.04912	−0.17904
10	Atlanta	2004	35.05321	34.7318	−1.81663	0.3851469
11	Baltimore	2004	35.0481	32.40218	−0.62902	−6.752003
12	Buffalo	2004	33.38227	35.437	4.590843	−3.461994
13	Carolina	2004	33.36462	31.58158	−0.78782	−0.139578
14	Chicago	2004	34.29233	32.21925	−8.48534	−0.253422
15	Cincinnati	2004	32.16609	32.99449	4.267261	1.6020636
16	Cleveland	2004	24.21647	43.11217	−1.50054	1.8877894
17	Dallas	2004	34.33833	33.25931	−2.97843	4.785603
18	Denver	2004	32.69762	28.89683	1.586123	−4.2706
19	Detroit	2004	33.34044	31.03221	−3.77983	1.3842011

Figure 46.1. Offense and defense salaries, and team performance data.

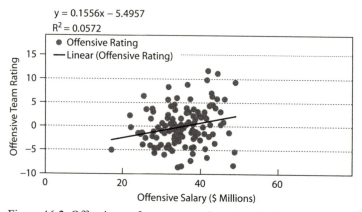

Figure 46.2. Offensive performance as a function of offensive team salary. See file NFLsalaries.xls.

implies that spending $1/.1556 = \$6.42$ million will improve offensive team performance by 1 point. The standard error of our regression predictions is 3.97 points. Any team whose predicted offensive performance is off by more than $2 \times 3.97 = 7.94$ points is considered an outlier or unusual observation. This is because in most cases around 95% of predictions from a regression are accurate within 2 standard errors. Notable among the outliers were the 2003 and 2004 Kansas City Chiefs, who outperformed our forecast by around 9 points. These outliers were mostly due to the great performance (20+ touchdowns each year) of Priest Holmes.

In a similar fashion (after adjusting the 2001–3 salaries so they are comparable to 2004 salaries), we can use the Excel Trend Curve to predict

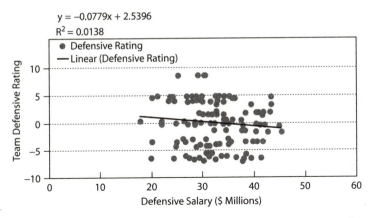

Figure 46.3. Defensive performance as a function of defensive team salary.

a team's defensive rating from their defensive players' salaries. The results are shown in figure 46.3.

We find that defensive team salary explains only 1.38% of the variation in team defensive performance. There is an −11.7% correlation between defensive team rating and defensive team salary. Remember that a negative defensive rating is better, so the negative correlation indicates that higher defensive salaries result in slightly better defenses. The weaker correlation for this regression indicates that NFL teams have more trouble identifying defensive talent than offensive talent. Our prediction for defensive team rating is defensive team rating = 2.5396 − .0779(defensive team salary in millions of dollars). This equation indicates that to improve a team by 1 point defensively, you need to spend 1/.0779 = $12.83 million.

In summary, for the recent NFL there is not a strong link between payroll and performance. It appears as if spending $1 million extra on offense will have a slightly better payoff (0.16 points of improvement) than spending an extra $1 million on defense (0.08 points of improvement).

Does a Larger Payroll Buy More Success in the NBA?

As our measure of NBA team performance we will use each team's season-ending Sagarin power ratings. We attempt to predict each NBA team's rating as a function of their payroll (using seasons 2005–7). A sample of the data is shown in figure 46.4.

	D	E	F	G
1			**RSQ** 0.012409	**CORR** −0.1114
2	**Year**	**Team**	**Salary**	**Rating**
3	2007	Atlanta	45.44	85.5
4	2007	Boston	62.22	85.1
5	2007	Charlotte	36.34	86.36
6	2007	Chicago	53.61	93
7	2007	Cleveland	63.05	93.84
8	2007	Dallas	94.96	96.96
9	2007	Denver	66.44	91.74
10	2007	Detroit	57.61	93.97
11	2007	Golden State	65.89	91.57
12	2007	Houston	62.91	93.79
13	2007	Indiana	69.61	87.56
14	2007	L.A. Clippers	57.33	89.8
15	2007	L.A. Lakers	77.42	90.47
16	2007	Memphis	64.19	84.91
17	2007	Miami	65.45	89.1
18	2007	Milwaukee	61.45	85.5
19	2007	Minnesota	65.64	87.07
20	2007	New Jersey	67.07	90.61
21	2007	New Orleans	65.02	89.54
22	2007	Orlando	59.84	88.79
23	2007	Philadelphia	96.17	87.57
24	2007	Phoenix	65.81	96.92
25	2007	Portland	77.42	86.98
26	2007	Sacramento	63.21	87.92
27	2007	San Antonio	66.69	98.54
28	2007	Seattle	57.37	87.31
29	2007	Toronto	53.99	90.5
30	2007	Utah	61.48	93.69
31	2007	Washington	62.54	88.52

Figure 46.4. NBA salaries and performance data, all teams included.
See file NBAteamsalaries.xls.

As we did in our NFL analysis, we adjusted the 2005 and 2006 salaries to be comparable to 2007 salaries. We found that only 1.2% of the variation in team rating is explained by variation in team salaries. There is a −0.11 correlation between salary and team performance, which seems to indicate that higher payrolls lead to worse performance. The Knicks are clearly an outlier. In each of the years under consideration the Knicks had the largest payroll and performed at a below-average level. We decided to delete the Knicks and rerun our model. The results are shown in figure 46.5.

After excluding the Knicks we find that only 0.09% of the variation in team performance is explained by variation in salary. The correlation between team performance and team salary is still negative (−.029).

	D	E	F	G
			RSQ	**CORR**
1			0.000884	−0.02973
2	Year	Team	Salary	Rating
3	2007	Atlanta	45.44	85.5
4	2007	Boston	62.22	85.1
5	2007	Charlotte	36.34	86.36
6	2007	Chicago	53.61	93
7	2007	Cleveland	63.05	93.84
8	2007	Dallas	94.96	96.96
9	2007	Denver	66.44	91.74
10	2007	Detroit	57.61	93.97
11	2007	Golden State	65.89	91.57
12	2007	Houston	62.91	93.79
13	2007	Indiana	69.61	87.56
14	2007	L. A. Clippers	57.33	89.8
15	2007	L. A. Lakers	77.42	90.47
16	2007	Memphis	64.19	84.91
17	2007	Miami	65.45	89.1
18	2007	Milwaukee	61.45	85.5
19	2007	Minnesota	65.64	87.07
20	2007	New Jersey	67.07	90.61
21	2007	New Orleans	65.02	89.54
22	2007	Orlando	59.84	88.79
23	2007	Philadelphia	96.17	87.57
24	2007	Phoenix	65.81	96.92
25	2007	Portland	77.42	86.98
26	2007	Sacramento	63.21	87.92
27	2007	San Antonio	66.69	98.54
28	2007	Seattle	57.37	87.31
29	2007	Toronto	53.99	90.5
30	2007	Utah	61.48	93.69
31	2007	Washington	62.54	88.52

Figure 46.5. NBA salaries and performance data, Knicks excluded.
See file NBAteamsalaries.xls.

The moral of our NBA analysis is that the size of a team's payroll has little effect on a team's performance. One reason that NFL payrolls seem to have a larger influence on team performance than do NBA payrolls is the fact that in the NFL no player's salary is guaranteed. You may cut an NFL player and the team does not have to pay him. This means that if an NFL team signs an expensive free agent and he does not pan out, the team can get his salary off the payroll. In contrast, most NBA contracts are structured such that an NBA team must pay a player even if they cut the player. So if an NBA team commits $100 million to a free agent and he does not fit in or is injured, the team can cut the player but they still have to pay him. In short, front office errors in judgment are much harder to fix in the NBA than in the NFL. If NBA teams are falling victim to the Winner's Curse (as explained in chapter 27), then they will have a much harder time recovering from their mistakes than will an NFL team. Once an NBA team blows $100 million on an overrated player (as when the Pacers signed Jalen Rose to a big contract), they do not have that money to spend on other players and their performance will surely suffer. The fact that NBA front office mistakes are harder to fix is consistent with our observation that extra dollars spent by an NBA team tend to be less effective than extra dollars spent by an NFL team.

Does a Higher Payroll Buy Increased Performance in Major League Baseball?

In our NFL analysis we used offensive and defensive ratings to represent team performance. In our NBA analysis we used Sagarin power ratings to measure team performance. In baseball it is difficult to develop a power rating for team performance because a team's performance is strongly tied to the ability of the starting pitcher. For this reason we will simply measure a baseball team's performance by their winning percentage. The file Baseballsalaries.xls contains MLB team payrolls and team winning percentages for the 2005, 2006, and 2007 (through August 30) seasons. A sample of the data is in figure 46.6.

We found that team salary explains 26.4% of the variation in team winning percentage. This is a much larger percentage than we found for the NFL or NBA. We speculate that since baseball is primarily an individual contest between pitcher and hitter, it is much easier to evaluate talent than it is to do so in football or basketball. The fact that the quality of information about baseball players' abilities is better than the information about football or basketball players' abilities would explain the fact that when baseball teams pay more for players, they are much more likely to get better players than is an NFL or NBA team.

Since we are trying to predict a team's winning percentage, we should use logistic regression instead of ordinary regression (recall in chapter 20 our attempt to predict the probability of a successful field goal as a function of the kick's distance). We found that our predicted team winning percentage based on team salary is as follows:

$$\text{predicted team winning percentage} = \frac{e^{-2.9232 + .003768 \times \text{salary}}}{1 + e^{-2.9232 + .003768 \times \text{salary}}}.$$

2	Team	Year	Record	Salary
3	Boston	2006	86–76	$120,100,524.00
4	L.A. Angels	2006	89–73	$103,625,333.00
5	Seattle	2006	78–84	$87,924,500.00
6	N.Y. Yankees	2006	97–65	$100,901,085.00
7	Arizona	2006	76–86	$58,884,226.00
8	Cleveland	2006	78–84	$56,795,867.00
9	N.Y. Mets	2006	97–65	$198,662,180.00
10	San Diego	2006	88–74	$68,897,179.00
11	Detroit	2006	95–67	$82,302,069.00
12	Philadelphia	2006	85–77	$87,148,333.00
13	Colorado	2006	76–86	$40,791,000.00
14	L.A. Dodgers	2006	88–74	$99,176,950.00
15	Minnesota	2006	96–66	$63,810,048.00
16	Chicago Cubs	2006	66–96	$94,841,166.00
17	Atlanta	2006	79–83	$92,461,852.00
18	Milwaukee	2006	75–87	$50,540,000.00
19	Toronto	2006	87–75	$71,915,000.00
20	St. Louis	2006	83–78	$78,491,217.00
21	Oakland	2006	93–69	$62,322,054.00

Figure 46.6. Major League Baseball team salaries and wins.

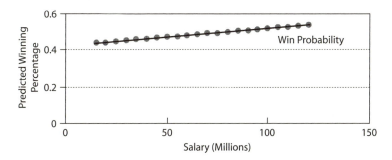

Figure 46.7. Team predicted winning percentage as a function of team payroll.

Here salary is measured in millions of dollars. Figure 46.7 shows the predicted winning percentage as a function of team salary. Every $10 million increase in team payroll increases a team's predicted winning percentage by around 1% (or 1.62 wins per season).

In summary, NBA team salaries are virtually uncorrelated with team performance. There is a weak correlation between NFL salaries paid to offensive and defensive players and offensive and defensive performance. The correlation is a little stronger for offensive salary and offensive team performance. There is a relatively strong relationship between baseball team payroll and baseball team winning percentage. Baseball teams probably get a better return for increased payroll because for baseball (primarily due to sabermetrics) there are better measures of player value than exist for football or basketball.

DOES JOEY CRAWFORD HATE THE SPURS?

In March 2007 NBA official Joey Crawford ejected Spurs star Tim Duncan during a game against the Dallas Mavericks. The NBA then suspended Crawford for the rest of the 2007 season. They said that he "was unfair to the Spurs." Is it possible to determine whether Joey Crawford's officiating harmed the Spurs' performance during Spurs games in which he officiated?

If Joey Crawford was biased against the Spurs, then we would expect the Spurs to have played significantly worse than expected during the games in which Crawford officiated. To determine an expected level of performance by the Spurs, we looked at each Spurs game officiated by Crawford during the 2004–5, 2005–6, and 2006–7 seasons. Using the season-ending Sagarin ratings and a home edge of 3 points we made a prediction for the amount by which the Spurs should have won each game. Then we computed for each game

$$\text{residual Spurs performance} = \text{actual Spurs winning margin}$$
$$- \text{predicted Spurs winning margin.}$$

For example, on April 15, 2007, the Spurs played the Mavericks at Dallas. They lost by 5 points. The Spurs had a season-ending rating of $+8.54$, while the Mavericks had a season-ending rating of $+6.96$. Our prediction for this game would have been that the Spurs would win by $(8.54 - 6.96) - 3 = -1.42$ points, or the Mavericks win by 1.42 points. Therefore, the residual Spurs performance for this game $= -5 - (-1.42) = -3.58$ points.

Historically the standard deviation of game margins about predictions made using the season-ending Sagarin ratings is around 12 points. If Crawford adversely affected the Spurs' performance, then we would expect the average of the Spurs' residuals to be significantly negative.

Our null hypothesis is that Joey Crawford's officiating has no effect on the Spurs. Our alternative hypothesis is that his officiating has a significant effect on the Spurs' performance (see chapter 11 for a discussion of hypothesis testing.) Each Spurs game residual is normally distributed and has a mean of 0 and a standard deviation of 12. The average of n identically distributed independent random variables has a mean equal to the individual random variables and a standard deviation equal to

$$\frac{(\text{standard deviation of individual random variable})}{\sqrt{\text{number of random variables}}}.$$

Our sample of Spurs games in which Crawford officiated includes 14 games. Therefore if Crawford has no effect on the performance of the Spurs, we would expect our average Spurs residual to be normally distributed with a mean of 0 and a standard deviation of

$$\frac{12}{\sqrt{14}} = 3.21 \text{ points.}$$

In reality our average Spurs residual was -2.5. This is $-2.5/3.21 = .78$ standard deviations below average. As we noted during our discussion of streakiness in chapter 11, any result more than two standard deviations away from expected causes us to reject our null hypothesis. Therefore, our analysis indicates that Crawford's officiating did not have a significantly adverse impact on the Spurs' performance.

The NBA could easily monitor the performance of each team in games involving each official and use our methodology to quell (or substantiate) teams' complaints about biased officiating.

DOES FATIGUE MAKE COWARDS OF US ALL?

The Case of NBA Back-to-Back Games and NFL Bye Weeks

"Fatigue makes cowards of us all" is a famous anonymous quote popularized by the late, great Green Bay Packers coach Vince Lombardi. The idea, of course, is that if you are tired you cannot perform at peak performance level. In this chapter we use the following two types of game situations to show that fatigue does indeed have a significantly deleterious impact on team performance:

- NBA teams that play back-to-back games perform significantly worse than expected during the second game of the back-to-back and perform even worse when they play a fourth game in five nights.
- The week after an NFL team has a bye or an open date they perform significantly better than expected.

NBA Back-to-Backs or Four Games in Five Nights

Typically an NBA team will play 10–20 games for which they also have played the night before. These games are called "back-to-backs." Back-to-back games are usually played in different cities. For example, the Dallas Mavericks might play Friday in Minnesota and Saturday have a home game. The NBA never schedules a team to play three consecutive nights.

One would think the combination of travel and having played a game the night before would lead to fatigue, which would cause inferior performance. Several times a year teams also play four games in five nights. Of course, the fourth game is a back-to-back game, but we would conjecture that the occurrence of two sets of back-to-back games in five nights would lead to even worse performance than a back-to-back game that was not the fourth game in five nights.

Using the game scores for the 2005–6 NBA season, let's take a look at the effects of back-to-back games and playing four games in five nights on team performances by following our "Joey Crawford" approach described in chapter 47. (See Backtobacks.xls.) For each game in which only one of the teams played a back-to-back game, we predicted the outcome of the game based on Sagarin ratings and home court edge as back-to-back margin prediction = (rating of back-to-back team) − (rating of non-back-to-back team) + (home edge of +3 if back-to-back team is home or −3 points if back-to-back team is away). Then we define for each game back-to-back residual = actual margin for back-to-back team − back-to-back prediction.

As described in chapter 47, if there are no back-to-back games and back-to-back games have no effect, then the average of these residuals should follow a normal random variable with mean of 0 and a standard deviation of $12/\sqrt{n}$. During the 2005–6 season, there were 314 back-to-back games that were not the fourth game in five nights and in which only one team faced a back-to-back. We found the sum of back-to-back residuals = −634.8. Thus, on average, the teams facing a back-to-back in this sample performed 634.8/314 = 2.02 points worse than expected. This is $\dfrac{-2.02}{\dfrac{12}{\sqrt{314}}} = -2.99$ standard deviations worse than expected. If back-to-back games have no effect on team performance, there are fewer than 2 chances in 1,000 of a discrepancy at least as large as −2.02 points. Therefore, our best estimate is that playing a back-to-back game (that is not the fourth game in five nights) when the opponent did not play the previous night results in a statistically significant decrease in performance of 2 points.

There were 56 games in which one team was playing the fourth game in five nights and the other team was not playing a back-to-back. In these 56 games the sum of the residuals was −228.1 points. This is an average of 228.1/56 = 4.1 points worse than expected. This is $\dfrac{-4.01}{\dfrac{12}{\sqrt{56}}} = -2.54$ standard deviations worse than expected. If the fourth game in five nights has no effect on team performance, there are fewer than 6 chances in 1,000

of a discrepancy at least as large as -4.01 points. Therefore, our best estimate is that playing a fourth game in five nights when the opponent did
not play the previous night results in a statistically significant decrease in
performance of around 4 points per game.

Does a Bye Week Help or Hurt an NFL Team?

During the NFL regular season each team gets a week off ("bye" week).
Let's take a look at whether teams play better or worse than expected the
week after the bye. Most people believe the week off favors the bye team
because the extra week of rest enables injured players to heal and gives the
coaches an extra week to prepare for their opponent. Others believe teams
get "stale" or "rusty" when they have an extra week off. Let the data decide
who is right. Our data consist of every post–bye week NFL game during
the 2003–7 regular seasons. We eliminated games in which both teams
were coming off a bye week, leaving us) with 146 games in which only one
team had a bye (see file Byeweek2.xls. As before, we computed bye team
margin prediction = (rating bye team) − (rating of non-bye team) + (home
edge of $+3$ if bye team is home or -3 points if bye team is away). Then for
each game, we define bye team residual = bye team actual margin − bye
team prediction. Over 146 games the sum of the bye week team residuals
was 381.1. Thus, on average, the bye week teams played 386.1/146 = 2.61
points better than expected. Since the standard deviation of the actual margin about a prediction based on the Sagarin ratings is around 14 points,
bye week teams played $\dfrac{2.61}{\dfrac{14}{\sqrt{146}}} = 2.25$ standard deviations better than we
would have expected if the bye week had no effect. The chance of a devia-
tion this large or larger is only 1 in 100. Therefore, we conclude that after
a bye week NFL teams do play significantly better (around 2.61 points
game better) than expected.

49

CAN THE BOWL CHAMPIONSHIP
SERIES BE SAVED?

As every college football fan knows, since 1998 the Bowl Championship Series (BCS) has selected two college football teams to play for the national championship in early January. In this chapter we will explain how the BCS currently (2007 season) ranks teams and chooses the two teams that play for the championship. We will also discuss two commonly suggested alternatives to the BCS: an eight-team playoff or a "plus-one" system that chooses the two teams that get to play for the championship after the New Year's Day bowl games.

A Brief History of the BCS

Starting in 1997, teams were ranked using the following four factors: subjective polls, computer rankings, strength of schedule, and team record.[1] In the first BCS Championship game on January 4, 1998, Tennessee defeated Florida State 23–16. In 2001 a "quality wins" factor that gave teams credit for defeating one of the top fifteen ranked teams was added to the mix of factors. During the 2001 season Nebraska made the championship game and was clobbered 37–14 by Miami. Most observers felt that Nebraska undeservedly made the championship game because many lopsided Cornhusker wins against weak teams "padded" their computer rankings. Therefore, beginning with the 2002 season, BCS computer ranking systems excluded the margin of victory from their algorithms. In 2004 the team records, strength of schedule, and quality wins were eliminated from the rankings because the BCS believed the computer rankings already included these factors. Let's take a look at the current (2007) ranking system.

[1] See http://www.bcsfootball.org/bcsfb/history for a complete history of the BCS.

The BCS 2007 Rankings

The BCS computer rankings for the 2007 season were computed by giving equal weight to three factors:

- The Harris Poll rankings: 114 former players, coaches, administrators, and media vote each week for the top twenty-five teams.
- The *USA Today* Coaches Poll: 60 current coaches vote each week for their top twenty-five teams.
- Computer rankings: six computer ranking systems are included. Most of the systems use ranking systems approximating the logit rating system described in chapter 40. (Recall that our logit methodology only counted wins and losses and ignored margin of victory.)
- The spreadsheet Bcsstandings.xlsx re-creates the computation of the BCS standings at the conclusion of the 2007 regular season (see figure 49.1).

Let's walk through how Ohio State received their .959 rating. For the Harris Poll each team receives 25 points for a first-place vote, 24 points for a second-place vote, 23 points for a third-place vote, and so on, down to 1 point for a twenty-fifth-place vote. From cell F4 we see that Ohio State earned 2,813 points in the Harris Poll. The maximum number of possible Harris points is $25 \times 114 = 2,850$. Therefore, the Buckeyes receive a Harris Poll ranking of $2,813/2,850 = .987$.

Similarly in the *USA Today* poll each team receives 25 points for a first-place vote, 24 points for a second-place vote, 23 points for a third-place vote, and so on, down to 1 point for a twenty-fifth place vote. Ohio State received a total of 1,469 points in the *USA Today* poll. Since a maximum of 60×25 *USA Today* poll points are possible, Ohio State receives a *USA Today* poll ranking of $1,469/1,500 = .979$.

To create a composite computer ranking from the six computer ranking systems, BCS again assigns for each computer system a score of 25 points for a #1 ranking, 24 points for a #1 ranking, and so on, down to 1 point for a #25 ranking. As is done in the judging of international diving and figure skating, the highest and lowest rankings are dropped. This helps avoid an "outlier" computer ranking from exerting undue influence on the final computer composite ranking. A team can obtain a maximum of $4 \times 25 = 100$ points from the four remaining computer point scores. Therefore their computer percentage is the sum of the remaining four computer ranking points divided by 100. Ohio State earned 25, 25, 21, 23, 22, and 21 points from the six BCS computers. After dropping the low score of 21 and the high

D	E	F	G	H	I	J	K	L	M	N	O	P	Q	R	S
	Harris Poll			USA Today			Computer Rankings						Comp. %	Comp. aver.	BCS aver.
Team	Rank	Points	%	Rank	Points	%	A&H	RB	CM	KM	JS	PW			
Ohio State	1	2813	0.99	1	1469	0.979	25	25	21	23	22	21	0.91	22.75	0.96
LSU	2	2630	0.92	2	1418	0.945	21	24	25	24	24	23	0.95	23.75	0.94
Virginia Tech	3	2345	0.82	3	1242	0.828	22	22	24	25	25	25	0.96	24	0.87
Oklahoma	4	2520	0.88	4	1331	0.887	18	21	18	18	23	24	0.8	20	0.86
Georgia	5	2469	0.87	5	1277	0.851	20	17	23	22	19	19	0.8	20	0.84
Missouri	6	2117	0.74	6	1104	0.736	24	16	22	21	20	22	0.85	21.25	0.78
USC	7	2346	0.82	7	1227	0.818	17	23	14	14	17	17	0.65	16.25	0.76
Kansas	8	2092	0.73	8	1099	0.733	23	20	16	20	21	20	0.81	20.25	0.76
West Virginia	9	1924	0.68	9	1010	0.673	16	15	20	13	15	18	0.64	16	0.66
Hawaii	10	1903	0.67	10	994	0.663	14	18	1	16	18	13	0.61	15.25	0.62
Arizona State	11	1628	0.57	11	900	0.6	19	19	17	17	13	16	0.69	17.25	0.62
Florida	12	1786	0.63	12	890	0.593	15	14	19	19	14	14	0.62	15.5	0.61
Illinois	13	1400	0.49	13	747	0.498	13	13	7	9	8	9	0.39	9.75	0.46
Boston College	14	1124	0.39	14	617	0.411	12	12	15	15	16	15	0.57	0.57	0.46
Clemson	16	1041	0.37	16	567	0.378	7	3	11	12	12	11	0.41	0.41	0.38
Tennessee	19	870	0.31	18	480	0.32	11	6	9	11	9	5	0.35	0.35	0.33
BYU	18	912	0.32	19	462	0.308	8	7	10	2	3	10	0.28	0.28	0.3
Wisconsin	15	1079	0.38	15	594	0.396	3	11	0	5	0	2	0.1	0.1	0.29
Texas	17	983	0.34	17	498	0.332	5	0	4	0	0	4	0.08	0.08	0.25
Virginia	21	551	0.19	20	332	0.221	6	2	8	7	10	8	0.29	0.29	0.23
South Florida	24	362	0.13	25	115	0.077	10	0	13	10	11	12	0.43	0.43	0.21
Cincinnati	20	580	0.2	23	215	0.143	1	5	6	4	4	6	0.19	0.19	0.18
Auburn	23	448	0.16	21	289	0.193	2	0	2	8	7	1	0.12	0.12	0.16
Boise State	22	541	0.19	22	246	0.164	0	10	0	0	0	0	0	0	0.12
UConn	29	52	0.02	T−28	23	0.015	0	8	3	3	6	7	0.19	4.75	0.07

Figure 49.1. BCS 2007 final season (pre-bowl) rankings. See Bcsstandings.xlsx. A&H = Anderson and Hester; RB = Richard Billingsley; CM = Colley Matrix; KM = Kenneth Massey; JS = Jeff Sagarin; PW = Peter Wolfe.

score of 25, we find that Ohio State's computer percentage is $(25 + 21 + 23 + 22)/100 = .91$. Now Ohio State's final BCS average is simply the average of the Harris Poll, *USA Today* Poll, and computer percentages $(.987 + .979 + .91)/3 = .959$. LSU ranked second with a BCS average of .939, so LSU played Ohio State in the championship game.

Note that the BCS averages three different rankings to compute an overall rating. The rankings are, in theory, good forecasters of future team performance (that is, performance in postseason bowl games). Economists have known for a long time that a forecast created by averaging the forecasts created from different forecasting methods usually is usually a more accurate predictor of the future than is each individual forecast.[2] Therefore

---

[2] See C.W.J. Granger, *Forecasting in Business and Economics* (Academic Press, 1989).

the BCS is applying sound economic theory by averaging three different forecasts. This logic assumes, however, that each forecasting scheme is accurate. Even in today's world of DVRs and TIVOs, the average poll voter probably cannot watch more than ten games a week. This makes it difficult for even the most dedicated poll voter to claim she is using all available information when she casts her vote each week. The BCS computers do "see every game" because all Division 1A game results are included in the computer rankings. Unfortunately, not allowing the computers to use the actual scores of the games discards much useful information. While realizing that the BCS does not want college football powerhouses running up the score on traditional opponents, it would seem that allowing the computers to utilize a margin of victory up to, say, thirty points would allow the computers to better rank the teams.

How Accurate Are the Polls and Computer Rankings? Why Do They Drive Fans and Sportswriters Nuts?

In NCAA basketball a 64-team tournament chooses the champion. Most fans believe the team that triumphs during March Madness is the best team at the end of the season and deserves the championship. Since the BCS picks two teams, there is plenty of room for argument about whether the best team brings home the trophy. For example, during the 2004 season Auburn, USC, and Oklahoma all went undefeated but only two teams (USC and Oklahoma) played for the championship. Auburn fans complained bitterly about being left out of the title game. During 2007 Ohio State ended the season as the only one-loss team. Out of a plethora of two-loss teams, LSU was chosen and they slaughtered the BCS's #1 ranked Buckeyes. As the father of a USC Trojan student ("Fight On"), I will make the case that USC (who demolished Illinois in the Rose Bowl) deserved a shot in 2007. Early in the season USC had many injuries to their offensive line. During their amazing upset at the hands of Stanford (who deserves a lot of credit) USC quarterback John David Booty played the whole game with a broken finger (unknown to the coaching staff at the time) and threw four interceptions.[3] He also sat out USC's other loss to a tough Oregon team. Using the Sagarin 2007 ratings we find that when USC was healthy during their

---

[3] For a video of the Stanford game, see http://stanford.fandome.com/video/86705/USC-Rewind-vs-Stanford-Pt-3/.

last four regular season games they played at a Sagarin rating level of 97 points. Over the same time period Ohio State played at a 93 level and LSU played at an 88 level. This indicates that at the end of the season USC was surely competitive with the two teams chosen by the BCS. For their first eight regular season games USC was crippled by injuries and played around 10 points worse than they did during their last four games. USC never got the chance to show what they could do against LSU.

Two suggestions are often proposed as an alternative to the BCS:

- an eight-team playoff involving the top eight teams in the BCS rankings
- a "plus-one" system in which the New Year's Day bowl games are seeded (#8 plays #1, #2 plays #7, etc.); the polls and computers are updated after the games and the #1 and #2 ranked teams play for the championship

If the goal of the BCS is to minimize the complaints that the "best team is not the champion," then an eight-team playoff is the best solution. The winner of an eight-team playoff will have defeated three excellent teams on their road to the championship. College football executives often say that an eight-team playoff would extend the season and make too many demands on student athletes. They fail to mention, however, that NCAA Division 1AA has a sixteen-team playoff involving student athletes from excellent schools such as William and Mary and Lafayette. We believe the real reason a playoff has not materialized is that it would reduce the importance of the major bowl games (Orange, Sugar, Cotton, Rose, and Fiesta), and the powerful bowl game lobby has denied the nation the tournament it deserves. Of course, an eight-team tournament would leave the ninth-ranked team crying foul because they would be denied their shot at the title.

Table 49.1 shows the difference in Sagarin point ratings between the season-ending best team in power ratings and the season-ending ninth-ranked team during 1998–2007. Thus in none of the ten years was the ninth-ranked team within five points in ability of the top-ranked team. It would seem the ninth-ranked team would have little reason to complain about exclusion from the playoff.

How about a "plus-one" playoff where two teams play for the title after the New Year's Day bowl games? We have already established that if there are eight spots open for the New Year's Day bowls, then the ninth-ranked team has little justification for complaining about not being included. If we choose the two "best" teams to play for the championship after the New Year's Day bowl games, what is the chance that the third-ranked team has a legitimate complaint about not being included? Table 49.2 gives the

TABLE 49.1
Difference in Sagarin Points between Number 1 and
Number 9 Teams in Ending Sagarin Ratings

Year	Difference
2007	6
2006	8
2005	20
2004	15
2003	9
2002	8
2001	19
2000	15
1999	11
1998	10

TABLE 49.2
Difference in Sagarin Points between Number 1 and
Number 3 Teams in Ending Sagarin Ratings

Year	Difference
2007	1
2006	1
2005	11
2004	11
2003	5
2002	2
2001	13
2000	5
1999	6
1998	2

difference in season-ending Sagarin power ratings between the top-ranked and third-ranked teams. During four of these ten years the third-ranked team was no more than 2 points worse than the top-ranked team and thus would certainly have had a legitimate complaint about not being allowed to play for the title.

In summary, an eight-team playoff would save the BCS. The current system will often lead to second-guessing about whether the best team won.

COMPARING PLAYERS FROM
DIFFERENT ERAS

In chapter 15 we tried to determine whether it would be likely that Ted Williams would hit .400 if he were to play today. Our analysis required that we compared the pitching and fielding abilities of players from different eras. In chapter 15 we used a fairly simplistic approach and found that it was unlikely that Ted Williams would hit .400 today. In this chapter we use our WINVAL ratings to determine whether the players in the NBA have improved or declined in quality since 2000. The end of the chapter summarizes the results of Berry, Reese, and Larkey, who analyzed the change in player quality over time for Major League Baseball, professional hockey, and professional golf.[1]

Analyzing Change in NBA Player Quality, 2000–7

We have WINVAL player ratings for all NBA players for the 2000–7 seasons. For example, if Dirk Nowitzki had a +10 rating for the 2004–5 season, that would indicate that per 48 minutes, if Nowitzki played instead of an average 2004–5 player, then our best estimate is that the team would improve their performance by 10 points per 48 minutes. We can use our WINVAL ratings to estimate the relative level of player abilities during the 2000–7 seasons. Let's arbitrarily assign the 2006–7 season a strength level of 0. If our model estimates, for example, that the 2003–4 season has a strength level of +4, that would mean, on average, players in 2003–4 were 4 points better than players in 2006–7. Each "data point" is a player's WINVAL rating for a given season. We restricted our analysis to players who played at least 1,000 minutes during a season. We used the Excel Solver to estimate:

---

[1] Berry, Reese, and Larkey, "Bridging Different Eras in Sports."

- for each player, an overall ability level relative to 2006–7 players (e.g., if we come up with a +10 estimate for Nowitzki, during our years of data he averaged playing 10 points better per 48 minutes than an average 2006–7 player); and
- an overall ability level for each season (e.g., if we obtain an estimate of −3 for the 2002–3 season, we would estimate players during the 2002–3 season were 3 points worse than 2006–7 players per 48 minutes).

Let's continue using Dirk Nowitzki as our example. We would hope that during each year

$$\begin{aligned}
&(\text{Dirk's ability relative to 2006–7 season}) \\
&= (\text{Dirk's rating during year x}) \\
&\quad + (\text{year x strength relative to 2006–7}).
\end{aligned} \tag{1}$$

For example, if Dirk had a +10 rating during 2002–3 and players during the 2002–3 season averaged 5 points better in ability than players during the 2006–7 season, then we would estimate that relative to 2006–7 players Dirk was 15 points better (per 48 minutes) than an average 2006–7 player. Rearranging equation (1) we can obtain

$$\begin{aligned}
&(\text{Dirk's rating during year x}) \\
&= (\text{Dirk's ability relative to 2006–7 season}) \\
&\quad - (\text{year x strength relative to 2006–7 season}).
\end{aligned} \tag{2}$$

We know the left side of (2) but we do not know either of the quantities on the right-hand side of (2). Following our approach to rating teams in chapter 40, we use the Excel Solver to choose each player's rating relative to 2006–7 and each season's average player rating relative to 2006–7 to minimize the sum over all players and seasons.

$$(\text{player rating during year x}) - \{(\text{player's ability relative to } \\ 2006–7 \text{ season}) - (\text{year x strength relative to 2006–7 season})\})^2.$$

After running this optimization we can find the estimates of player strength for each season relative to 2006–7. For example, the average level of player ability in the 2000–2001 season was 0.78 points worse than an average 2006–7 player. For example, the 2005–6 season player strength was virtually indistinguishable from that of the 2006–7 season, as illustrated in table 50.1.

Twenty years from now we should have about thirty years of player rating data. This will enable us to settle the debates about whether the NBA

TABLE 50.1
Season Strengths Relative to 2006–7

Season	Relative Player Strength
1999–2000	−0.32
2000–2001	−0.78
2001–2	−0.56
2002–3	−0.69
2003–4	−0.17
2004–5	−0.53
2005–6	−0.02
2006–7	0

stars of the 2020s are better than LeBron James, Kobe Bryant, and Dirk Nowitzki.

Bridging Eras in Sports: A More Sophisticated Approach

Berry, Reese, and Larkey have analyzed the changes over time in abilities of National Hockey League (NHL) players, golfers, and MLB players.[2] Their major goal was to compare the abilities of players from different eras. They used 1996 as their "base" season. They developed equations similar to (1) and (2) for their analysis but also included for each player some additional terms to measure the influence of a player's age on the player's performance. The sections that follow summarize their results.

Aging in Hockey, Golf, and Baseball

BRL found that hockey players improve steadily in their ability to score points (points = assists + goals) until age 27 and then experience a sharp decline in ability. They found that golfers improve until age 30–34, with there being little difference in a golfer's ability in the 30–34 age range. For baseball, BRL found that home run–hitting ability increases until age 29

[2] Ibid.

and then drops off. Batting ability peaks at age 27 and then drops off, but not as steeply as hockey ability drops off post-peak.

Comparing the All-Time Greats

BRL found that Mario Lemieux and Wayne Gretzky were the two greatest hockey players (non-goalies) of all time. BRL estimated that at their physical peak, Mario Lemieux would be a 187-point player and Wayne Gretzky would be a 181-point player. They determined that Gordie Howe was the best "old-timer," estimating him to be a 119-point player. Again, these numbers are predictions for how the player would perform in the base year (1996).

BRL determined that Jack Nicklaus, Tom Watson, and Ben Hogan were the three best golfers of all time. For example, BRL projected that in a Grand Slam tournament in 1996, Nicklaus (at his peak) would average 70.42 strokes per round, Watson (at his peak) would average 70.72 shots per round, while Hogan (at his peak) would average 71.12 shots per round. Tiger Woods was just beginning his legendary career and BRL estimated that Woods in his peak would average 71.77 strokes per round in a 1997 Grand Slam tournament. In hindsight, it is clear that this estimate underestimated Woods's abilities.

BRL estimate that the best all-time hitter for average was the legendary Ty Cobb. They estimate that if he were at his peak in 1996 he would have hit .368. Second on the list was Hall of Famer Tony Gwynn. If at his peak in 1996, Gwynn was estimated to be a .363 hitter. Finally, Mark McGwire was estimated to be the best home run hitter of all time. BRL estimated that if McGwire was at his peak in 1996 he would have averaged 0.104 home runs per at bat. Second on the list was Texas Ranger Juan Gonzales, who was projected to hit 0.098 home runs per at bat during 1996 if he had been at his peak. The legendary Babe Ruth was estimated to hit 0.094 home runs per at bat in 1996 if he were at his peak. BRL did their study before Barry Bonds's home run hitting really took off, so they estimate that if at his peak Bonds would have hit only 0.079 home runs per at bat in 1996. BRL's work shows how a sophisticated mathematical model can answer age-old questions such as who is the greatest home run hitter of all time?

51

CONCLUSIONS

We have covered a lot of material in this book. We have shown how applying mathematics can improve the performance of baseball, football, and basketball teams. We have learned how to evaluate teams and players and determine the probabilities of interesting events such as consecutive no-hitters. We have also gained an understanding of how sports gambling works. In this chapter we will review the important mathematical tools that we have used in our analysis.

The Use of Regression

Throughout the book, we have used regression to try to understand how various team statistics impact team performance. Often we tried to predict a dependent variable (such as Runs Scored, or an NFL team's points for and against) from team statistics (hitting statistics, team offense and defense statistics in the NFL). We used regression to analyze several sports situations.

- We showed how the Linear Weights (see chapter 3) derived by multiple regression can be used to predict how singles, walks, doubles, triples, and home runs contribute to team Runs Scored and runs generated by an individual.
- In our study of football we used regression (chapter 18) to derive the surprising result that an NFL team's passing yards per attempt on offense and defense explain nearly 70% of the team's overall performance.
- In our study of NBA basketball (see chapter 28) we found that a team's offensive and defensive effective shooting percentage were much more important to determining overall team performance than were rebounding, turnovers, and free throw efficiency.
- We saw that NBA referees are more likely to call fouls on players of a different race (chapter 34) by taking a look at Price and Wolfers's regression analysis.[1]

Analyzing Key Game Decisions

In analyzing game decisions we often developed a payoff measure (e.g., probability of winning a game in baseball or basketball, expected runs in an inning, expected points by which we outscore the other team in football) and determined which decision maximizes the expected payoff. Some examples of this methodology are summarized below.

- Not bunting yields more expected runs per inning than bunting (see chapter 6).
- Going for it on fourth down and short yardage often yields a better change in the expected margin of victory than does a field goal or punt; thus teams should be less conservative on fourth down (see chapter 21).
- Passing on first down leads to a better change in expected margin than does running; thus teams should pass a little more (but not so much that defense can adjust to this change in strategy) (see chapter 21).
- Going for a three-point shot at the end of regulation when a team is down by two points is a better option than going for two points because going for three maximizes the chance of winning (see chapter 37).

Evaluate Players Based on How They Help Their Team Win

What matters is winning the game, so players should be evaluated based on how they increase or decrease their team's chance of winning a game. Some of these techniques are summarized below.

- We used SAGWIN points in baseball to analyze how each batter faced by a pitcher or batter plate appearance changed his team's chance of winning (see chapter 8).
- For the 2006 Colts (see chapter 22), we computed the average number of points gained or lost for all "skill positions" (quarterback, running back, or receiver) by the player's team on plays involving the player. We found, for example, that Joseph Addai was a much better runner than Dominic Rhodes. Of course, we do not know how much a running back's performance is due to his offensive line, and we do not know how much a quarterback's performance is due to the offensive line and the quality of the team's receivers.

---

[1] Price and Wolfers, "Racial Discrimination among NBA Referees."

- In basketball (see chapter 30) we looked at how game margin or chance of winning changes when a specific player is on the court, and apportioned this change among players. A player's performance can depend on his teammates on the court and opponent matchups.
- We can translate how much (see chapters 9 and 33) a baseball player or basketball player helps his team into his fair salary by linking SAGWINDIFF (in baseball) or WINVAL (in basketball) to find his value created over replacement player (VORPP).

Ranking Teams

Based on the scores of football, soccer, and basketball games it is easy to rank teams and determine the strength of the team's schedule.

- We can rank teams (see chapter 40) by simply finding the set of ratings and home field edge that best predict each home team margin of victory by (home edge) + home rating − away rating.
- A team's schedule strength is simply the average ability of the team's opponents.
- Ratings for baseball teams would be misleading because the starting pitcher has an important impact on which team wins the game.
- Organizations such as the BCS in college football (see chapter 49), the NCAA in college basketball, and FIFA in World Cup Soccer often compromise the fairness of their "tournaments" by using inferior quantitative metrics. In college basketball the RPI (see chapter 42) is a fatally flawed indicator of a team's ability because it is logically inconsistent and ignores game scores. This probably leads to "bubble teams" being unfairly slighted by the NCAA tournament committee. In college football the BCS computers are hamstrung by not being allowed to use margin of victory to rate teams. This may lead to the two best teams not playing for the championship.

Hypothesis Testing Lets Us Understand the World

We can use hypothesis testing to determine whether a factor has a statistically significant effect on a quantity of interest. Simply collect data and predict a base case for the data under the assumption that our factor does not matter. If what we observe is more than 2 sigma away from what is expected in the base case, then what we have observed has less than a 5% chance of happening. We then conclude our factor is important. Here are some examples in which we used hypothesis testing to gain insights into sports phenomena of interest:

- We used a runs test (chapter 11) to show that streakiness rarely exists for baseball hitters or NBA team performance.
- We determined (chapter 48) that NBA teams performed 2 sigma worse than expected (based on our least squares ratings) during back-to-back games. We also showed that after a bye week NFL teams performed more than 2 sigma better than expected.
- We found that after adjusting for other referees and the propensity of teams committing and drawing fouls that during games in which Tim Donaghy officiated and the Total Line moved at least 2 points, the number of free throws attempted was 2 standard deviations more than expected (chapter 36).
- We used hypothesis testing to show that during games he officiated, Joey Crawford did not significantly affect the performance of the San Antonio Spurs (chapter 47).

Basic Probability Theory and Monte Carlo Simulation Can Be Used to Compute Many Quantities of Interest

Throughout the book we have used basic probability theory (the theory of independent events, the normal random variable, and the Poisson random variable) to compute many probabilities that interest sports fans including the following:

- the probability of Joe DiMaggio's 56-game streak or consecutive no-hitters ever occurring (see chapter 16);
- the probability of many great sports collapses, such as the collapse of the 2007 Mets or 1964 Phillies (chapter 45);
- the probability that a team wins a game or covers a bet (chapter 43); and
- the probability of teams triumphing in multigame tournaments such as the NBA Finals or NCAA tournament.

The Conventional Wisdom Is Often Wrong

For years baseball fans thought fielding percentage was the right way to measure a fielder's effectiveness. As we saw in chapter 7, Bill James and John Dewan showed that the conventional wisdom was wrong. In chapter

19 we saw that proper analysis provides an easily understood method to rank NFL quarterbacks that is superior to the NFL's current, incomprehensible method. In chapter 21 we showed that NFL teams should be less risk averse and eschew a punt or a field goal on fourth down much more often than they actually do.

Life Is Not Fair

Former president Jimmy Carter once said that "Life is not fair." We have seen that the BCS and NCAA basketball use unfair methods to select teams for the BCS Championship and to seed and select teams for the NCAA basketball tournament. In chapters 25 and 26 we saw that neither the NFL's nor the NCAA's football overtime procedure is fair. We saw that a simple "cake-cutting" procedure would make football overtime fair in the sense that each team would think they have at least a 50% chance to win.

In chapter 40 we found that the huge difference in quality of the eight World Cup groups in 2006 allowed inferior teams to advance to the later rounds at the expense of superior teams that were placed in stronger groups. This distasteful situation could have easily been avoided if better computer rankings were used to seed the groups.

Life may not be fair but in sports, math can help level the playing field and ensure that the outcome of important contests is fairly determined.

A Final Word

As is sung in the *Lion King*, "There is more to see than can ever be seen and more to do than can ever be done." I feel this way about math and sports. There are many unsolved important problems in sports for which mathematical analysis can (I hope) provide a solution. For example, are wide receivers or the offensive line more important to the passing game? In which situations is running a better call than passing? I hope this book will inspire and equip readers to join the mathletics revolution.

LIST OF DATABASES

The letters *t* or *f* following a page number indicate a table or figure on that page.

ANNOTATED BIBLIOGRAPHY

This is not a complete bibliography but rather a road map that will help guide the reader through the vast mathletics literature.

Books

Adler, Joseph. *Baseball Hacks*. O'Reilly Media, 2006.

This great book provides a one-stop entry into the field of sabermetrics. For the reader with a computer programming background, this book provides the tools to use the baseball data available on the Internet to perform almost any imaginable analysis.

Baseball Prospectus Team of Experts, Jonah Keri, and James Click. *Baseball between the Numbers: Why Everything You Know about the Game Is Wrong*. Perseus Publishing, 2006.

This book contains many advanced sabermetric essays dealing with topics including the following: Does clutch hitting exist? How do we compare players from different eras? The glossary of sabermetric terms alone is worth the price of the book.

Berri, David, Martin Schmidt, and Stacey Brook. *The Wages of Wins: Taking Measure of the Many Myths in Modern Sport*. Stanford University Press, 2006.

The authors develop simple yet effective metrics for ranking football quarterbacks and running backs. Most of the book is devoted to developing the Win Score and Wins Produced metrics that are used to evaluate NBA players (see chapter 29). David Berri's Web site, http://dberri.wordpress.com/, contains many posts about the NBA.

Cook, Earnshaw. *Percentage Baseball*. MIT Press, 1966.

Earnshaw Cook's pioneering work sparked a great deal of controversy when it first appeared. Cook tried to link runs scored and given up to number of games won, thereby anticipating the Pythagorean Theorem discussed in chapter 1. Cook also showed that bunting is usually a bad idea (see chapter 6).

Dewan, John. *The Fielding Bible*. Acta Sports, 2006.

This book revolutionized (as described in chapter 7) the methods used to analyze players' fielding abilities. The book is no longer updated annually but the *Bill James Handbook* provides many of Dewan's valuable fielding statistics.

Gennaro, Vince. *Diamond Dollars*. Maple Street Press, 2005.

Vince Gennaro provides details regarding how a baseball team's market size and ability (are they in the playoff mix?) affect the amount a team should pay for a player. The book also discusses in great detail how to build a perennial contender.

Goldman, Steven. *Mind Game: How the Boston Red Sox Got Smart, Won a World Series, and Created a New Blueprint for Winning*. Workman Publishing, 2005.

    This is an entertaining read that explains how rigorous analysis helped the Red Sox make many of the key decisions that led to the long-awaited 2004 World Championship.

Goldman, Steven, and Christina Kahrl. *The Baseball Prospectus 2008*. Plume Publishing, 2008.

    Much of the important work in sabermetrics has been published on the Baseball Prospectus Web site (see below). This "almanac" is updated annually and contains lots of interesting advanced baseball statistics.

James, Bill. *The Bill James Handbook 2008*. ACTA Sports, 2007.

    Updated annually, the handbook contains many useful advanced baseball statistics such as Park Factors (discussed in chapter 10) and John Dewan's fielding ratings (discussed in chapter 7).

James, Bill. *The New Bill James Historical Baseball Abstract*. Rev. ed. The Free Press, 2001.

    This is a revised edition of Bill James's 1985 classic.

James, Bill. *Win Shares*. STATS Publishing, 2002.

    In this book Bill James explains his complicated method for determining how many of a baseball team's wins can be attributed to each player. I prefer the simpler method based on Player Win Averages (see chapter 8).

Levitt, Steven. *Freakonomics: A Rogue Economist Explores the Hidden Side of Everything*. William Morrow, 2005.

    Steven Levitt masterfully describes how clever and sophisticated techniques used by today's economists enable them to obtain surprising insights into human behavior.

Lewis, Michael. *Moneyball*. W. W. Norton, 2004.

    This is a great read that describes how the Oakland Athletics spent far less money on player salaries than did teams like the New York Yankees but still performed well. The secret, of course, was using data to determine an objective value for players and draft picks. Michael Lewis is one of sabermetrics' best writers and this book is a true page turner.

Mills, Eldon, and Harlan Mills. *Player Win Averages*. A. S. Barnes, 1970.

    This book was the first to describe the vitally important (yet simple) idea that players should be evaluated on how they change the chances that their team wins a game (see chapter 8).

Oliver, Dean. *Basketball on Paper*. Potomac Books, 2003.

    Dean Oliver has worked for the Seattle Sonics and currently works for the Denver Nuggets as a statistical consultant. In this book Oliver shows how to use math to improve and evaluate a basketball team's performance. His brilliant four-factor model is discussed in chapter 28.

Palmer, Peter, and John Thorn. *The Hidden Game of Baseball*. Doubleday, 1985.

    This book covers a lot of ground. In particular, the authors do a great job of describing how Monte Carlo simulation (see chapter 4) can be used to evaluate batters.

Poundstone, William. *Fortune's Formula*. Hill and Wang, 2005.
> A page-turning history of the Kelly money management formula and many other topics in modern finance.

Schwarz, Alan. *The Numbers Game: Baseball's Lifelong Fascination with Statistics*. Thomas Dunne Books, 2004.
> This book is a well-written, highly readable history of the evolution of baseball statistics. The book contains great portraits of sabermetrics pioneers such as Bill James.

Schatz, Aaron. *Pro Football Prospectus 2008*. Plume, 2008.
> Each year Aaron Schatz and his staff at Footballoutsiders.com put out this great "football almanac." The book contains projections for team and player performances as well as much original research dealing with the football-math interface.

Shandler, Ron. *Ron Shandler's Baseball Forecaster 2008*. Shandler Enterprises, 2007.
> Each year Ron Shandler puts out his forecasts for how each major league player will perform the following season. Many fantasy baseball players think his forecasts are the best around, but to my knowledge no comparison has been made between the accuracy of Shandler's forecasts to those of other sabermetricians such as Bill James.

Tango, Tom, Mitchell Lichtman, and Andrew Dolphin. *The Book: Playing the Percentages in Baseball*. Potomac Books, 2007.
> Every chapter in this book is a model of impeccable mathematical analysis of an important topic (does platooning matter; when should teams bunt; and so forth). The writing is clear. If a sabermetrician is to be marooned on a desert island with one book, this should be it. Tom Tango's Web site, http://www.tangotiger.net/, also contains a lot of great sabermetric research.

Thorn, John. *Total Baseball*. 8th ed. Warner Books, 2004.
> This book is a statistical encyclopedia of baseball team, player, and manager statistics organized around *Washington Post* columnist Thomas Boswell's Total Baseball version of linear weights.

Von Neumann, John, and Oskar Morgenstern. *Theory of Games and Economic Behavior*. Princeton University Press, 1944.
> This is the book that started game theory.

Winston, W. *Data Analysis and Business Modeling with Excel 2007*. Microsoft Press, 2007.
> This book provides the tools needed to manipulate data in Excel and analyze it. If the chapter appendixes on Excel leave you thirsting for more Excel knowledge, then this is the book for you. For example, chapter 5 on text functions shows many ways to modify data imported from the Internet or a database.

Wong, Stanford. *Sharp Sports Betting*. Pi Yee Press, 2001.
> This is the single best book for an introduction to sports betting.

Newspaper, Magazine, Journal, and Online Articles

Alamar, Benjamin. "The Passing Premium Puzzle." *Journal of Quantitative Analysis in Sports* 2, no. 4 (2006): article 5.

Alamar provides strong evidence that NFL teams run more often than they should on first down (see chapter 21).

Albright, S. C. "A Statistical Analysis of Hitting Streaks in Baseball." *Journal of the American Statistical Association* 88 (1993): 1175–83.

S. C. Albright shows that contrary to popular perception, very few hitters earn the label "streak hitter" (see chapter 11).

Annis, David. "Optimal End-Game Strategy in Basketball." *Journal of Quantitative Analysis in Sports* 2, no. 2 (2006): article 1. http://www.bepress.com/jqas/vol2/iss2/1.

This is an interesting article that uses decision trees to analyze whether a team with a three-point lead should foul the opposition near the end of the game.

Berry, Scott, S. Reese, and P. Larkey. "Bridging Different Eras in Sports." *Journal of the American Statistical Association* 94 (1999): 661–76.

This is a remarkable essay that shows how the power of statistics enables us to compare the abilities of golfers, baseball players, football players, and hockey players even if their lives or playing careers do not overlap.

Bialik, Carl. "Should the Outcome of a Coin Flip Mean So Much in NFL Overtime?" *Wall Street Journal*, December 23, 2003. http://online.wsj.com/article/SB107152932067874700.html.

This article discusses several interesting methods to make overtime more equitable, including bidding for the ball, moving the kickoff, dueling kickoffs, and giving both teams the ball (see chapter 26).

Carter, Virgil, and Robert Machol. "Operations Research on Football." *Operations Research* 19, no. 2 (1971): 541–44.

This is the first attempt to create the yard-line values for first down and 10 yards to go that are discussed in chapter 20. (Virgil Carter was a starting quarterback for the Cincinnati Bengals.)

Falk, R. "The Perception of Randomness." In *Proceedings of the Fifth Conference of the International Group for the Psychology of Mathematics Education*. Laboratoire IMAG, 1981.

This article shows how people have difficulty differentiating between random and non-random sequences.

Gilovich, T., R. Vallone, and A. Tversky. "The Hot Hand in Basketball: On the Misperception of Random Sequences." *Cognitive Psychology* 17 (1985): 295–314.

This classic work (discussed in chapter 11) dispels the myth that basketball players' shooting exhibits a "hot hand."

Gould, Steven. "The Streak of Streaks." *New York Review of Books* 35, no. 13 (1988). http://www.nybooks.com/articles/4337.

This is a beautifully written article that passionately makes the case that Joe DiMaggio's 56-game hitting streak (discussed in chapter 16) is the greatest sports record of all time.

Heston, Steve L., and Dan Bernhardt. "Point Shaving in College Basketball: A Cautionary Tale for Forensic Economics." Working paper, 2006. http://ssrn.com/abstract=1002691.

This is a brilliant refutation of Justin Wolfers's claim in "Point Shaving in College Basketball" that point shaving exists in college basketball.

Kalist, David E., and Stephen J. Spurr. "Baseball Errors." *Journal of Quantitative Analysis in Sports* 2, no. 4 (2006): article 3. http://www.bepress.com/jqas/vol2/iss4/3.

This article discusses many matters of interest concerning the official scorer's decisions to call a batted ball a hit or an error.

Kelly, L. "A New Interpretation of Information Rate." *Bell System Technical Journal* 35 (1956): 917–26.

This classic article describes the Kelly growth criteria used for optimal money management in gambling and investing. (See chapter 44 for a discussion of the use of the Kelly criteria for determining optimal bet size in sports gambling.)

Kubatko, Justin, Dean Oliver, Kevin Pelton, and Dan T. Rosenbaum. "A Starting Point for Analyzing Basketball Statistics." *Journal of Quantitative Analysis in Sports* 3, no. 3 (2007): article 3. http://www.bepress.com/jqas/vol3/iss3/1.

This article contains many useful definitions of important basketball statistical terms.

Lawhorn, Adrian. " '3-D': Late-Game Defensive Strategy with a 3-Point Lead." http://www.82games.com/lawhorn.htm.

This is a discussion by statistician Adrian Lawhorn (although based on limited data) on whether an NBA team that is leading by three points at the end of a game should foul their opponent.

Leonhardt, David. "It's Not Where NBA Teams Draft, But Whom They Draft." *New York Times*, June 26, 2005. http://www.nytimes.com/2005/06/26/sports/basketball/26score.html.

This article discusses the efficiency of the NBA draft (see chapter 33).

Levitt, Steven. "Why Are Gambling Markets Organised So Differently from Financial Markets?" *Economic Journal* 114 (2004): 223–46.

This is a brilliant discussion by the author of the best-seller *Freakonomics* that shows why it is a good idea to bet on the point-spread underdog when betting on NFL games.

Lindsey, George. "An Investigation of Strategies in Baseball." *Operations Research* 11, no. 4 (1963): 477–501.

This article contains brilliant discussions of Linear Weights (see chapter 3) and baseball decision-making (see chapter 6).

Mease, David. "A Penalized Maximum Likelihood Approach for the Ranking of College Football Teams Independent of Victory Margins." *American Statistician* 57 (2003): 241–48.

This article shows how to rate teams when at least one team has 0 wins or at least one team has 0 losses (see chapter 40).

Price, Joseph, and Justin Wolfers. "Racial Discrimination among NBA Referees." Working paper, 2007. http://papers.ssrn.com/sol3/papers.cfm?abstract_id=1136694.

This essay (discussed in chapter 34) shows that when calling fouls, NBA referees have a slight (but statistically significant) tendency to favor players of the same race.

Romer, David. "It's Fourth Down and What Does the Bellman Equation Say?: A Dynamic-Programming Analysis of Football Strategy." February 2003. http://emlab.berkeley.edu/users/dromer/papers/nber9024.pdf.

> This is the controversial article by Berkeley economist David Romer later published in the *Journal of Political Economy* (April 2006), which uses dynamic programming to show that NFL coaches are much too conservative on their fourth-down decision-making. Many NFL coaches reacted to the article with skepticism. See http://espn.go.com/nfl/columns/garber_greg/1453717.html.

Rosen, Peter A., and Rick L. Wilson. "An Analysis of the Defense First Strategy in College Football Overtime Games." *Journal of Quantitative Analysis in Sports* 3, no. 2 (2007). http://works.bepress.com/peterrosen/1.

> This is the first article to document that college football teams that choose to kick rather than receive win more overtime games (see chapter 25).

Sackrowitz, Harold. "Refining the Point(s)-After-Touchdown Scenario." *Chance* 13, no. 3 (2000): 29–34. http://www.amstat.org/PUBLICATIONS/chance/pdfs/133 .sackrowitz.pdf.

> This is the first attempt to determine whether (based on the score of a game and the number of possessions remaining) a team should go for a one-point or two-point conversion. The exposition in chapter 24 is very similar to Harold Sackrowitz's approach.

Shor, Mike. "Game Theory and Business Strategy: Mixed Strategies in American Football." http://www2.owen.vanderbilt.edu/Mike.Shor/courses/game-theory/ docs/lecture05/Football.html.

> Mike Shor has put together many interesting illustrations of game theory including a football example (see chapter 23).

Smith, Michael David. "The Race of Truth: 40-Yard Times Can Tell the Future." *New York Times*, April 27, 2008. http://www.nytimes.com/2008/04/27/sports/ football/27score.html?_r=1&scp=1&sq=keeping+score&st=nyt&oref =slogin.

> This article (described in chapter 27) shows how a running back's weight and 40-yard dash time can be used to predict his future NFL success.

Stern, Hal. "On the Probability of Winning a Football Game." *American Statistician* 45 (1991): 179–83.

> This article shows that the standard deviation of NFL game scores about the point spread is around 14 points. This is used in chapter 43 to determine the probability of winning various types of bets on NFL games.

Thaler, Richard H., and Cade Massey. "The Loser's Curse: Overconfidence vs. Market Efficiency in the National Football League Draft." Working paper, 2006. http:// mba.yale.edu/faculty/pdf/massey_thaler_overconfidence_nfl_draft.pdf.

> As discussed in chapter 27, the authors purport to show that NFL draft selectors do a poor job of selecting players. On his blog, Phil Birnbaum has pointed out some serious flaws in the Thaler-Massey analysis (http://sabermetricresearch .blogspot.com/2006/12/do-nfl-teams-overvalue-high-draft-picks.html).

Wagenaar, W. "Generation of Random Sequences by Human Subjects: A Critical Survey of Literature." *Psychological Bulletin* 77 (1972): 65–72.

This article discusses how people have difficulty spotting random sequences.

Wolfers, Justin. "Point Shaving in College Basketball." *American Economic Review* 96, no. 2 (2006): 279–83.

Although published in a prestigious journal and cited in several books such as Ian Ayres's *Supercrunchers*, I believe the main argument of this article has been successfully refuted by Steve Heston and Dan Bernhardt in "Point Shaving in College Basketball."

Online Resources

http://www.82games.com

Roland Beech's Web site analyzes play-by-play data for each player and gives a tremendous amount of valuable information that can be used to analyze team and player strengths and weaknesses. For example, for the 2007–8 season, WINVAL found Jarred Jeffries was a near average NBA player. From the player stats section of Jeffries's page we find he shot 26.7% effective field goal percentage on jump shots. If he could improve on this abysmal performance he would be an above-average player. WINVAL had Eddie Curry ranked as one of the league's worst players during the 2007–8 season. Looking at the On Court/Off Court Stats page for Curry we find that when he is on the court the Knicks commit roughly 6 more turnovers per 100 possessions (there are approximately 100 possessions in one game) than their opponents and when Curry is off the court the Knicks commit roughly 1 fewer turnover per 100 possessions than do their opponents. This horrible turnover "inefficiency" explains in large part Curry's poor Adjusted + / − rating.

http://danagonistes.blogspot.com/

For years Dan Agonistes' Web site has provided intelligent commentary on sports and math (with the emphasis on baseball). The blog entry http://danagonistes.blogspot.com/2004/10/brief-history-of-run-estimation-runs .html contains an excellent summary of the historical development of the Runs Created concept.

http://www.baseball1.com/

This fantastic Web site provides access to Sean Lahman's amazing baseball database. You can download each player, team, or manager's statistics in a spreadsheet format. Each gives the relevant statistics for a given season. For example, row 5000 may give Babe Ruth's 1927 batting statistics. If you have Excel 2007 (which handles over 1,000,000 rows of data) you can quickly do great analysis with this data.

http://www.baseballprospectus.com/ and http://www.hardballtimes.com/

These two Web sites contain an incredibly comprehensive body of work on sabermetrics. For example, Voros McCracken's pioneering work on the unpredictability of pitching performance (discussed in chapter 5) and Keith Woolner's concept of Value Over Replacement Player Points (see chapter 9) were first published on the Baseball Prospectus Web site.

http://www.baseball-reference.com/

This is a great site for downloading a player's career statistics or a team's game-by-game scores for a season. For example, if you want Barry Bonds's year-by-year statistics in a spreadsheet simply copy the data and paste it into Excel. Each year's data will be pasted into a single cell, but with Excel's great Data Text to Column feature (see Winston, *Data Analysis*) you can easily parse the data so that each number is placed in its own cell.

http://www.basketball-reference.com

Justin Kuratko's outstanding Web site is the perfect source for team, player, and coaches' statistics. In particular, under Advanced Statistics you can find each player's PER rating.

http://www.bbnflstats.com/

Brian Burke's great blog on mathematical analysis of football has many excellent research studies involving the NFL. (See the study of quarterback ratings in chapter 19.)

http://www.bcsfootball.org/

This is the official Web site of the Bowl Championship Series (BCS). Be prepared to be deluged with content that extols the virtues of the BCS.

http://www.espn.com

ESPN's site contains detailed play-by-play descriptions for each baseball, football, and basketball game. For football these data are needed for the player and play effectiveness analysis discussed in chapter 22. For example, after selecting NFL from the home page you may select Schedule and then choose a year and then a week of the season and pull up any game's play-by-play going back to the 2002 season.

http://www.footballcommentary.com/

This Web site contains a lot of sophisticated NFL analysis. Most important, the authors use dynamic programming (the same approach used to analyze two-point conversions in chapter 24) to determine when teams should go for it on fourth down. These tables are located at http://www.footballcommentary.com/goforittables.htm.

http://www.footballoutsiders.com

This is the preeminent site for football analysis. This site contains sophisticated ratings for players and teams (see chapter 22). New research studies are also often posted.

http://www.goldsheet.com

If you want access to game results and point spreads this site is for you. The Gold Sheet is a gambling tip sheet and the Web site gives scores and point spreads for the NFL, NHL, NBA, MLB, college basketball, and college football. Copy and paste the data into Excel and then use text functions to clean it up. See chapter 5 of Winston, *Data Analysis*.

http://kenpom.com/rpi.php

This is the best site to learn about how college basketball's flawed RPI works. The site also updates the RPI ratings during the college basketball season.

http://www.kiva.net/~jsagarin/mills/seasons.htm

This site contains the Player Win Ratings for 1957–2006 (see chapter 8). It is fun to look at a player like Sandy Koufax and see his transformation from a below-average pitcher in 1957 to a Hall of Fame–level pitcher during the early 1960s.

http://www.nba.com

The official NBA site contains great information including complete play-by-plays for each game and Pure + / − numbers updated during a game. The Lenovo stats

(http://www.nba.com/statistics/lenovo/lenovo.jsp) let you view Pure +/ − numbers for players and two-man, three-man, four-man, and five-man combinations.

http://pigskinrevolution.com/aboutus.html

This is the home page for the Zeus computer program, which uses data (and dynamic programming) from many NFL seasons to recommend play calling strategy for fourth-down situations and after touchdown conversions.

http://www.pro-football-reference.com/

This is a great site that provides statistics for players, coaches, and teams. If you want to download a player's statistics using a different season in each row of a spreadsheet, this is the site for you.

http://www.retrosheet.org/

This is a great site that provides many game box scores and play-by-play accounts. If you read the book *Baseball Hacks*, you can do wonders with this data.

http://sabermetricresearch.blogspot.com/

This is perhaps the best mathletics blog on the Internet. Sabermetrician Phil Birnbaum gives his cogent reviews and analysis of the latest mathletics research in hockey, baseball, football, and basketball. This is a must-read that often gives you clear and accurate summaries of complex and long research papers.

http://sonicscentral.com/apbrmetrics/

This is the home page for the Association for Professional Basketball Research. Many people post comments on the use of mathematics to analyze basketball.

http://sportsmogul.com/

Clay Dreslough's Web site contains the DICE formula (see chapter 5) used to predict a pitcher's ERA from strikeouts, home runs, and walks. There are also several outstanding sports simulation games marketed from this site.

http://stat.wharton.upenn.edu/~stjensen/research/safe.html

This site contains the fielding statistics developed by Wharton professors. While John Dewan breaks the field into a small number of zones, Spatial Aggregate Fielding Evaluation (SAFE) in effect breaks the baseball field into an infinite number of zones and should theoretically result in a more accurate estimation of a player's fielding abilities.

http://www.usatoday.com/sports/sagarin.htm

The famous Sagarin ratings for the NBA, NHL, NFL, college football, and college basketball are located here. This Web site also includes Jeff Sagarin's Runs Created ratings for batters and pitchers based on Markov chain analysis. This methodology is similar to the Monte Carlo simulation method for evaluating hitters discussed in chapter 4. The Web site http://www.usatoday.com/sports/sagarin-archive.htm contains ratings from past seasons.

http://winexp.walkoffbalk.com/expectancy/search

This is a fantastic site that gives (based on games from the 1977–2006 seasons) the probability that a major league team will win a game given any score differential, inning, or on-base situation.

INDEX

The letters *t* or *f* following a page number indicate a table or figure on that page. For citations of databases, see the List of Databases (p. 341).